Geophysical and geotechnical methods for diagnosing flood protection dikes

Guide for implementation and interpretation

Cyrille FAUCHARD, Patrice MÉRIAUX

This guide was produced as part of the "CriTerre" National Project, a project set up by the French Civil and Urban Engineering Network (RGCU), subsidised by the Department of Scientific and Technical Research (DRAST) – which is part of the French Ministry for Infrastructure, Transport, Spatial Planning, Tourism and the Sea (METATTM) and coordinated by IREX, the Institute for applied Research and Experimentation in Civil Engineering.

The following have also contributed to the funding and/or management of this work: the LCPC* in Nantes, Cemagref in Aix-en-Provence, EDG (a private company), the Loire-Bretagne water agency (particularly, the Multi-disciplinary Team working on the "Loire management Plan") and the Agly river local manager.

* Laboratoire central des Ponts et Chaussées

Cyrille Fauchard (LRPC, Rouen) and Patrice Mériaux (Cemagref, Aix-en-Provence) would also like to thank:
– Gérard Degoutte – Cemagref, Aix-en-Provence – for his careful re-reading and invaluable assistance in producing this guide;
– the following persons for their re-reading: Alain Tabbagh and Roger Guérin (UPMC Paris 6), Richard Lagabrielle and Philippe Côte (LCPC, Nantes), Pierre Frappin (EDG);
– Paul Royet and Rémy Tourment (Cemagref, Aix-en-Provence);
– the following persons for discussions and providing documents: Alain Hollier-Larousse† (LCPC, Nantes), Viviane Borne (Calligée, Nantes), Bernard Bourgeois (BRGM, Orléans), Gregory Bièvre (LRPC, Autun), Olivier Magnin (Terraseis);
– the following bodies/institutions for the consultation or reproduction of documents: the CETMEF, the Isère-Drac-Romanche rivers regional manager, IMS (a private company based in Montbonnot), Saint-Nazaire University, Indre-et-Loire department of public works, the SAFEGE private office (Nanterre) and the LRPC in Saint-Brieuc.

Translation credits:
– Aquitaine Services Logistique, Lormont, France (www.asl.fr) – Richard Turner.
– Special thanks to Cathy Ricci, graduate in English, for translation review.

© 2004 Cemagref-Éditions, IREX, for the French guide. Cover photographs – Dikes on the Agly river: Panorama of the breach at St Laurent-la-Salanque (© Eric Josse DDE 66), Protection at the foot of the bank partially carried away by the flood of 12/13/11/1999 (© Patrice Mériaux).
Publishing coordinator: Julienne Baudel. Layout: Desk, 25 bd de la Vannerie 53940 Saint-Berthevin. Computer graphics: Françoise Cedra, Françoise Peyriguer, Joëlle Veltz. ISBN 978-2-7592-0031-3. 1st edition. Copyrighting Éditions Quæ: 2nd quarter 2007. Diffusion: www.quæ.com tel. +33 1 30 83 34 06. **Price: 40 €**

TABLE OF CONTENTS

Table of contents

1 Introduction

The flood protection dikes that run alongside many French rivers have been modified and upgraded many times over the years. Often very old, these works are mostly made of earth and have been built in stages. Details of their structure are not generally known, and they are usually heterogeneous, either in transverse section through the dike (*e.g.* a weighted zone in one of the slopes) or in the longitudinal profile (*e.g.* a repaired breach).

The spectacular and occasionally dramatic floods of the last decade both in Europe and in the United States have confirmed the chronic vulnerability of these dikes and the need for their diagnostic analysis. Moreover, review of breach repair work (in some cases of recent repairs) clearly shows that this work has often been conducted in haste, with insufficient control of the filler materials and using materials that do not match the original construction materials. Details of the repair work tend to fade quickly from archived records, with the result that heterogeneities lay hidden within the body of the dike – potential points of weakness for the structure that must first be detected using reliable methods, and then characterised during a subsequent diagnosis.

In this context, the investigative work conducted to diagnose dikes now tends to combine geophysical methods with more traditional geotechnical methods (test drills, *in situ tests*, etc.). Compared with geotechnical methods, geophysical methods generally offer the advantage of very short exploration increments (that only slightly, if at all, affect the efficiency); however, they only produce "apparent" and overall values of a particular soil property.

As a general rule, difficulties can arise when applying geophysical or geotechnical methods to dikes:

– they are "dry" (no hydraulic head) for most of the time and the critical element in their vulnerability at times of flood, *i.e.* seepage, is absent at the time of the investigation;

– their great length poses a crucial problem in finding a balance between cost and technical performance when conducting investigative work.

To address these and other issues, the LCPC (French Public Works Research Laboratory) in Nantes and Cemagref in Aix-en-Provence, France, conducted experimental research from 1998 to 2004 within the framework of the "CriTerre" National research Project ("Improvement of soils and control of reinforced soils" – theme "Detection of soil anomalies"), coordinated by the IREX (French Institute for applied Research and Experimentation in Civil Engineering). The aim of this research was to test and evaluate, for two French dike systems (one on the Cher river in the "Indre-et-Loire" department and the other on the Agly coastal river in the "Pyrénées-Orientales" department), the high-efficiency geophysical and geotechnical surveying methods applicable to dikes, and to draw from this experimental work the methodological elements required for the optimal utilisation of these tools.

This experimental research work forms the basis for this guide, which describes a general, three-phase approach to the high-efficiency diagnosis of "dry dikes"[1].

The first phase of the diagnosis, the "preliminary studies", is an essential prerequisite for the study. It involves collecting as much information as possible about the history of the dike (construction, breach history), its external characteristics (topography, how it is maintained, signs of deterioration), and its role in the local system (local geology and river dynamics). This information is collated in various ways – searching through archives and analysing them, interviews with the manager responsible for the various structures and a visual inspection. This study must determine the nature of the component materials of the dike, the nature of the foundation on which it rests and the hydraulic and morphodynamic conditions that it must

---

1. Synonymous with: "flood protection dikes" that do not have to resist a permanent hydraulic head; these dikes are generally built above the normal water level of the river and are "dry" when the diagnostic survey is performed.

resist. The quality of the final diagnosis is dependent on the thorough and rigorous execution of this phase of the study.

The main aim of **the second phase** of the study – the geophysical survey – is to isolate heterogeneous portions of the dike, and to determine those sections of the structure which, due to their differing physical characteristics, may be the point of initiation of irreversible damage (particularly breaches) during a flood. The geophysical methods used must satisfy, as a minimum, the following two main requirements: firstly the need to survey over long distances and secondly the need to identify the degree of heterogeneity of the structure over its entire height (including its foundation). The information collected, following correlation with the preliminary studies, is then used firstly to define specific zones that are the focus for localised investigation methods to determine, in detail, the geometry across a horizontal or transverse section of the dike, and secondly to set up the geotechnical surveys that constitute phase three.

This third phase comprises various geotechnical tests and drillings that ascertain *in situ* the principal mechanical characteristics of the materials that make up the structure. The results obtained are used to calibrate the geophysical measurements, and can lead to the deployment of further localised geophysical investigations.

These three steps, if conducted correctly, provide all the elements required for a thorough investigation of a dry dike. Since methodological tools have been in short supply in the field of <u>geophysical surveying</u>, (but not for the first and the third phases mentioned above, that are already covered in various French guides: cf. bibliography), the methodology proposed in this guide includes a comprehensive description of the principles underpinning the geophysical methods, their implementation and interpretation. The aim of this guide is to provide owners, project managers, contractors or others with all the elements they require to assess the strengths and weaknesses of the dikes diagnosed. This work can then serve as the basis for reinforcement work or for more in-depth studies.

Although the principles of the method described herein were developed on dikes in France, they may be applied with confidence to flood protection dikes and levees in other countries around the world. For this reason, the original French version of this guide (written in 2004) has been translated into English for this edition.

2

BACKGROUND TO THE CIVIL ENGINEERING DIAGNOSIS OF FLOOD PROTECTION DIKES

2.1 Nature, functions and composition of dikes

This section is a summary of chapter 1 of the guide produced by Mériaux *et al.* (2001).

2.1.1 Definition

A dike (or levee, in the context of this guide) is a linear structure that protects against flooding, with at least some of the structure rising above the flood plain. Its main function is to contain high water levels during floods and thus protect areas that are naturally prone to flooding. A typical cross-section through a dike is a berm of earth built up from the original ground with the protected zone[2] on one side and the river bed on the other. Depending on the circumstances, the dike is built either alongside the main channel, or set back from the river on the flood plain (figure 1).

---

2. Also called: [protected] valley.

Dikes are usually "dry" structures, insofar as hydraulic head is only a factor at times of flood; this is an important point when considering how they work and in selecting appropriate investigative techniques.

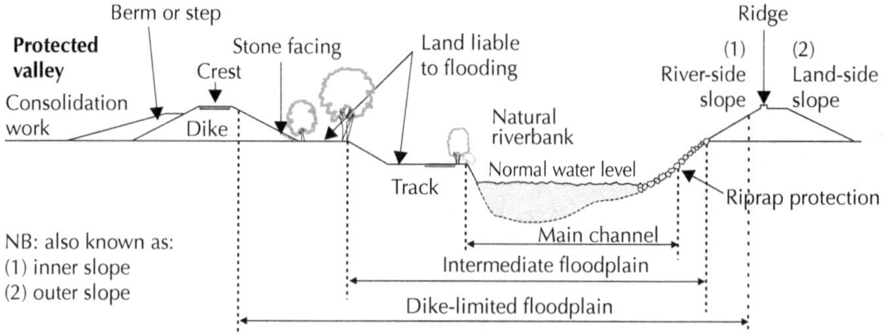

Figure 1 – Typical cross-section across a dike-protected valley (Mériaux *et al.*, 2001)

2.1.2 How dike systems work

Dike systems work as follows:

1. Rising water levels cause the river to break out of its main channel (figure 2) and spill out over the flood plain. The dike stops the flooding spreading to the protected valley.

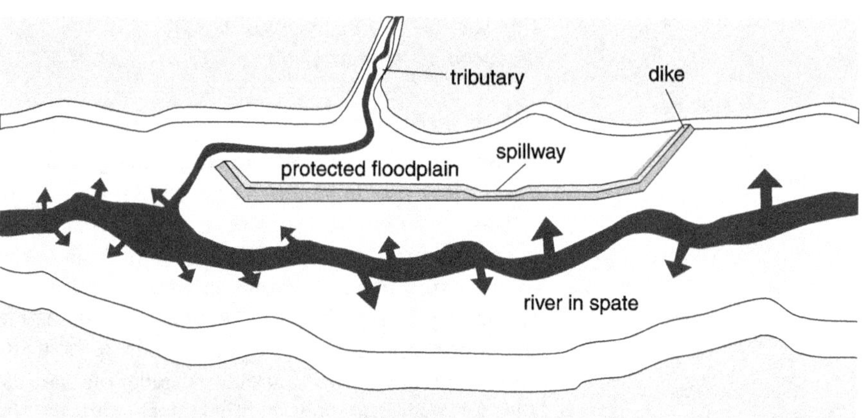

Figure 2 – Flood waters spreading into a dike-protected flood plain (Mériaux *et al.*, 2001)

2. The dike system is designed to limit the impact of low- and medium-severity floods on the valley that it protects. Inevitably, however, compared to the situation with no dikes, the contained water rises to a higher level, and especially if the diked channel is narrow (a common feature in urban areas).

3. In this scenario, the potential for flood peak reduction (relieving maximum down-stream flow rates by inundating the flood plain) is limited.

4. Even so, the protected zones may be flooded by a rise in the water table (figure 3), by runoff from a catchment basin, or even by the river backing up into a tributary.

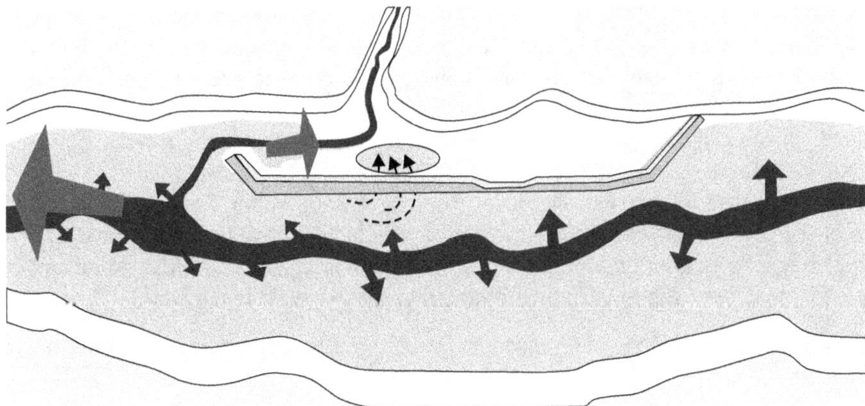

Figure 3 – Flooding of a valley by backing up, by runoff from a catchment basin or a rise in the water table (Mériaux *et al.*, 2001)

5. To reduce the risk of uncontrolled overtopping in cases of severe flooding, spillways (figure 4) are sometimes built into the dikes. These spillways can also be used to reduce downstream flow rates by flooding less critical valleys upstream.

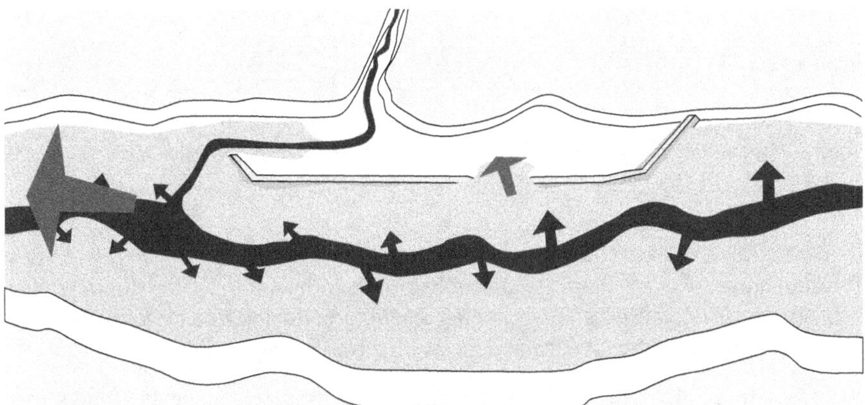

Figure 4 – Operation of a spillway (Mériaux *et al.*, 2001)

6. When river levels are exceptionally high the entire valley will flood, due either to the operation of a spillway, or to the effect of one or more breaches opening in the dike.

2.1.3 Composition of a dike

a) EARTHFILL DIKE

A dike is generally made of loose materials (sand, silt-loam, clay). Its morphological and mechanical characteristics vary greatly, depending on its geographical location, history, the water courses that it borders and on the floods that it is designed to contain.

Most French dikes were first constructed many years ago, and built using the resources and techniques available at the time. Consequently, the materials that make up the body of the dike were generally sourced from close to the structure and tend to be sandy in the middle reaches of rivers, and loamy nearer to the mouth. This approach produced poorly-compacted structures, insufficiently pervious and with no cut-off trench into the foundation. Moreover, they have been strengthened and/or raised over the years in response to the damage caused by historical floods (figure 5). Their heterogeneity is even greater, either through a transverse section (due to raising or broadening) or in the longitudinal profile (*e.g.* post-breach repair work).

Clearly dikes are highly heterogeneous structures, both in type and composition (figures 6 and 7).

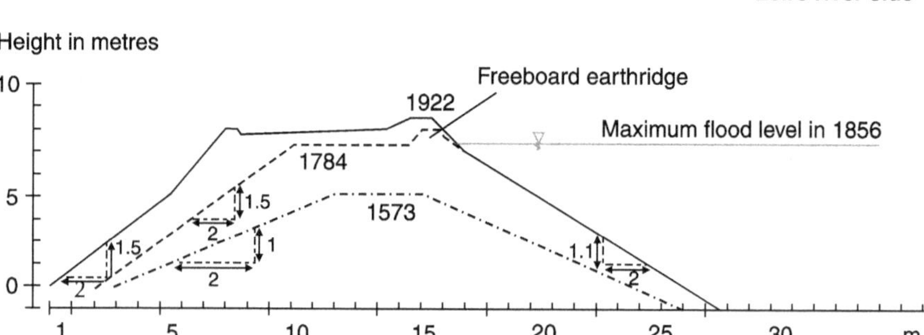

Figure 5 – Typical cross-section of a Loire river levee before recent strengthening work, showing the work conducted after historical floods (Dion, 1934)

b) THE IMMEDIATE ENVIRONMENT

The sloping surfaces of dikes are generally protected by grassing. On the river side, sections in contact with the main channel are often protected by a masonry pitching, in some cases covered by a natural deposit of silt or by vegetation.

In zones exposed to scouring, linear protective features, generally made from secant wood piles, have sometimes been used at the toe of the river-side slope of the dike.

Finally, in order to provide a freeboard that can let the dike resist a maximum design flood, the crest of the structure is often topped with a ridge of earth or a small masonry wall, called a "banquette" in French.

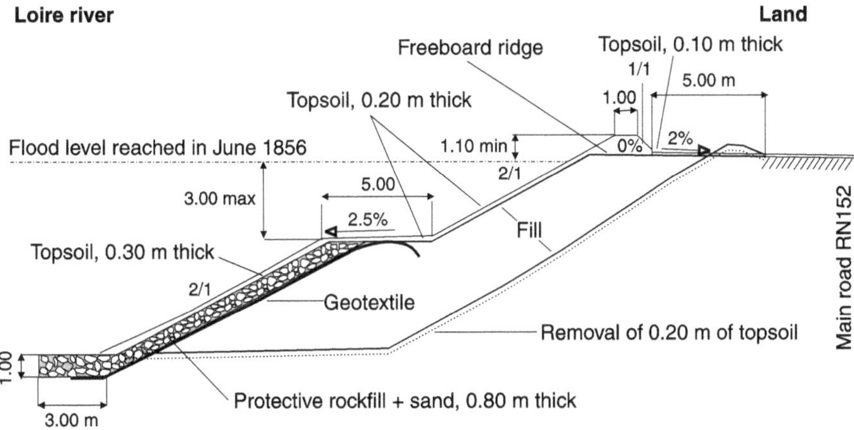

Figure 6 – Typical cross-section showing reinforcement on the Loire river side of the Fondettes-Luynes dike – DDE Project Indre-et-Loire (1997)

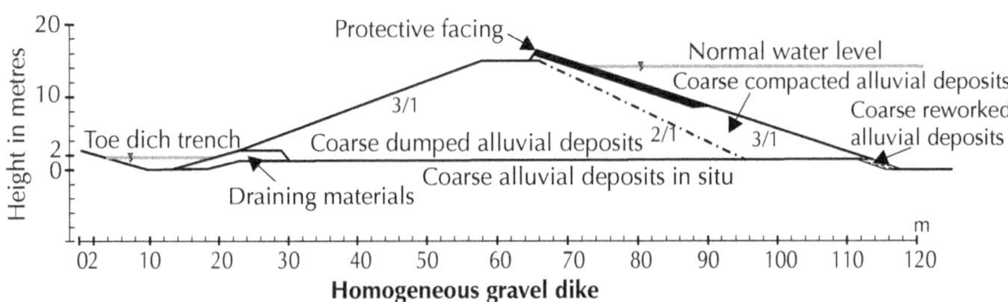

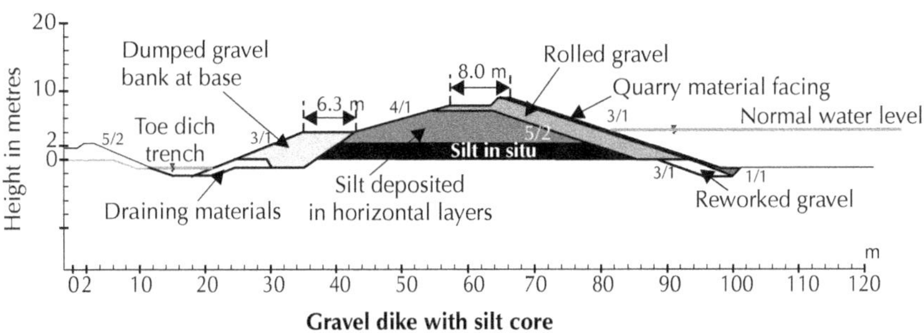

Figure 7 – Typical cross-sections of a dike for a hydroelectric development on the Rhone river (Mériaux et al., 2001, taken from a CNR diagram)

Currently, the design of this type of structure is very similar to that of small earthfill dams or of embankments for hydroelectric developments (figure 7). A soil mechanics study must be performed to determine the mechanical and sealing characteristics of the materials. The dike may be divided up into zones to separate the sealing function from the mechanical stability function. Particular attention must be paid to the problem of draining seepage water away from the dike, which may involve the use of draining materials. The reader may wish to refer to Degoutte (1997) for a description of construction principles for small dams.

This guide will not dwell on the design of new dikes, since its primary focus is on existing dikes and their diagnosis.

c) Quay walls made from masonry

When the available width for a structure is limited, which is often the case when a river passes through an urban area – "gravity retaining wall" type dikes are built. Older dikes are made from cut stone, whereas more recent examples are made of concrete. Their foundations may be on piles. On the land side, they may be supported by a shoulder of earth or rock, and in some cases are topped with a road.

d) Spillways

Spillways (figure 8) are designed to come into operation only during rare floods (typically to cope with hundred-year floods). They are normally a few dozen centimetres (typically 1 metre) lower than the crest of the dike. They spill water as the flood peaks into a less critical zone such as a flood meadow and thus prevent or delay an overtopping-type failure of the dike. Spillways are low points built into the longitudinal section of the dike. They usually consist of a weir, with a chute and then a stilling basin on the land side slope to control and dissipate the energy of the flowing water. The concrete or masonry sill is topped in some cases by a sand ridge (also called "banquette"), which is slightly lower than the crest of the dike. This "banquette" is a deliberate point of weakness that is eroded rapidly in the event of overtopping, and its function is to delay the operation of the spillway, so as to optimise its effect on peak water levels.

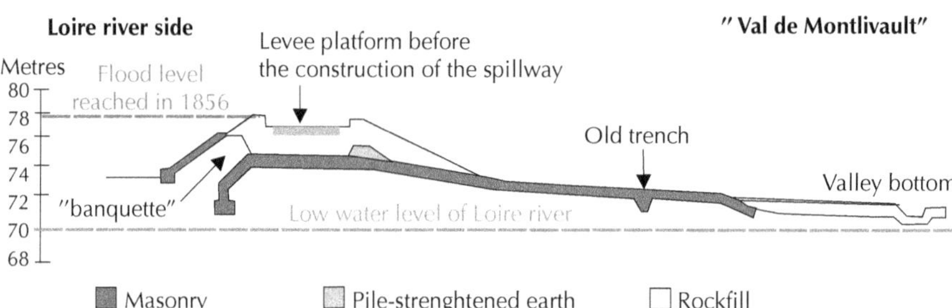

Figure 8 – Typical cross-section of an overflow weir in the Montlivault dike (Loir-et-Cher) (1890) (Lino *et al.*, 2000)

e) Particular structures

It is not unusual to find a variety of structures in dikes, often traversing and in many cases difficult to spot: buried channels, tunnels, aqueducts, various pipes and cables (hydraulics, electrical, telephone, etc.) travelling along or through the dike, inspection features, access ramps, stairs, road parapets, and sometimes even cellars or houses, as seen in the Loire valleys.

2.2 Classification of malfunctions and failure mechanisms

The remainder of this guide will focus primarily, and unless otherwise indicated, on "earthfill" type dikes.

The principal mechanisms that result in deterioration or a partial or total failure (breach) of a dike are:

– erosion of the surface of the dike (crest or land side slope) caused by overtopping, that very often results in a total breach;

– erosion on the river side due to the current, and particularly erosion at water level (scouring) that may result in hazardous landslides when the dike runs alongside a water course;

– internal erosion of the dike or of its foundation, that also very often results in a total breach, particularly if the flood is long-lasting;

– the slumping of one of the two slopes, and particularly the collapse of the slope on the river side in the event of a rapid drop in water level.

In rare cases, the force of the water pushing against the structure can also cause the dike to fail, especially where the dikes are narrow. This situation can occur following the collapse of one of the two slopes or after severe erosion of the slope on the river side.

2.2.1 Overtopping

Overtopping (figure 9) occurs when the water level rises above the crest of the dike.

It quickly creates a breach, often as a result of erosion initiated at the toe of the land side slope (if it is not submerged) that develops back up to the crest. This is the principal identified failure mechanism for earthfill dikes.

In the absence of protection, dikes are broadly acknowledged to be unable to resist overtopping.

Moreover:

– sandy materials and their heterogeneity tend to make a dike more vulnerable to breaching from overtopping;

– irregularities in the structure at the crest or on the land side slope (levelling faults, differential compaction, poor quality earthworking) induce a local concentration of overtopping flows, that accelerate the erosion;

– on the contrary, a well-compacted structure, with a uniform longitudinal profile, a gently-sloping slope, well grassed and a hard-surface crest, will resist a few centimetres of overtopping for a limited period of time.

Because dikes are submerged whenever the river reaches a certain level (during a rare flood), it is very important to know how often this event occurs, (recurrence interval of 10 years, 20 years or 100 years, etc.), so that the degree of protection provided by the structure is known.

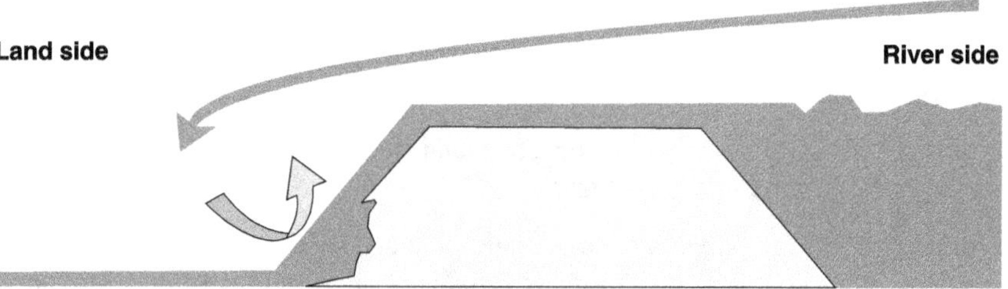

Figure 9 – Principle of overtopping (Mériaux *et al.*, 2001)

Returning-flow failures

This is a particular case of overtopping that occurs when the water that overflowed rejoins the river downstream and crosses the dike again. It also occurs when the discharge from a tributary basin saturates its outflows into the river and fills the valley. The breach occurs more rapidly if the structure had been soaked for a long time, and then the flow that rejoins the river further increases greatly the downstream flood.

2.2.2 External erosion and scouring on the river side

These mechanisms (figure 10) are caused by currents and/or waves, on the river side slope of the dike, or even on the banks running alongside the dikes. They are more powerful at times of flood and cause erosion first at the toe of the body of the levee.

The development of such erosion weakens the mechanical stability of the structure (steeper gradient or creation of overhangs). It can lead to earth slides, which in turn can initiate eddy and erosive phenomena that can accelerate the opening of a breach.

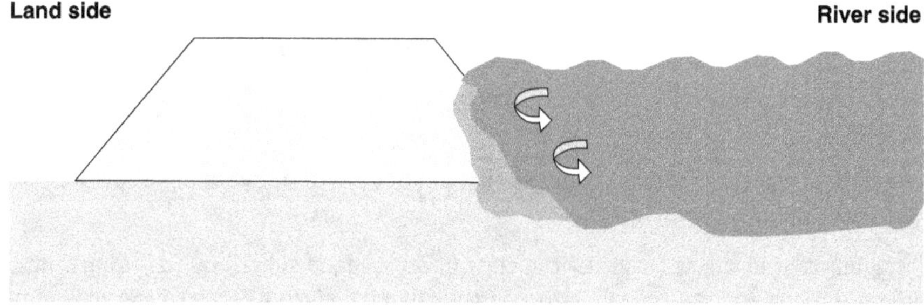

Figure 10 – Principle of external erosion and undercutting (Mériaux *et al.*, 2001)

2.2.3 Internal erosion (or piping)

During a flood, a hydraulic head is created that produces a hydraulic gradient through the body of the dike and its foundation. When it reaches a critical value, which varies depending on the nature of the materials, water begins to flow through specific zones and internal erosion phenomena appear. The result may be the creation of a tunnel (or pipe) through the structure, followed by a breach. On the surface, the internal erosion may be revealed by the appearance of sink holes.

Factors that exacerbate (figure 11) these phenomena are:

1. the presence of tunnels or excavations in the body or foundation of the structure (animal burrows, holes left by rotted tree roots, construction work in the body of the dike);

2. poor sealing between the body of the dike and the heterogeneities mentioned above (construction work, tunnels, etc.);

a) Natural factors

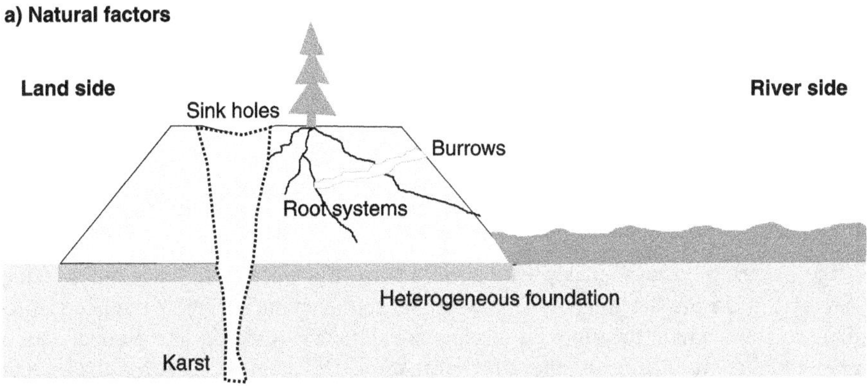

b) Anthropological factors

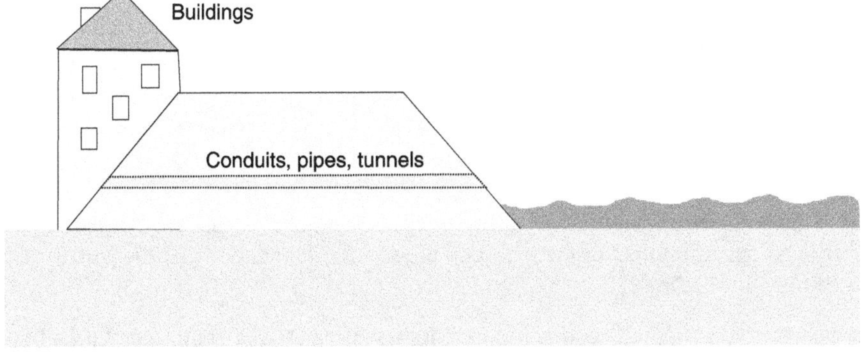

Figure 11 – Natural (*a*) and anthropogenic (*b*) factors exacerbating internal erosion (Mériaux *et al.*, 2001)

3. heterogeneities in the layers of materials that form the dike, and more seriously the foundation which in only rare cases is treated appropriately;

4. the relative permeability of the body of the dike overall. If the permeability of the dike materials is high, then its profile should be broad (*i.e.* the width-to-height ratio should be high) to avoid the risk of internal erosion.

Finally, large sink holes may appear on the surface of the dike (figure 11) or at its foot, resulting from water circulation and more rarely from collapses in a foundation that includes a karstic level.

2.2.4 Sliding of the slope

The land side slope may creep or even slide when the dike becomes saturated (during a sufficiently long flood). If this occurs when the dike is loaded, the sliding of the land side slope may cause the dike to fail (breach) with severe consequences. Three factors aggravate the sliding phenomenon:

– steep slopes (slopes greater than 2/3 or batters of less than 3H/2V);

– high pressures in the dike due to the absence of drainage and the presence of heterogeneous layers;

– weak mechanical properties of the fill materials (low density) or the presence of a poorly-consolidated clay bed in the foundation.

All three of these factors may be concomitant in zones where a breach occurred in the past.

If the water level drops quickly after a flood peak, the slope of the dike or its facing may also slide on the river side. This phenomenon, more common than land-side slides, occurs when the slope is steep and/or contains clayey material or has a sealed facing. This complex phenomenon requires the slope to be saturated and/or for uplifts to develop under the facing once the stabilising effect of the hydraulic head has disappeared on the river side. Sliding is more likely if the slopes have been eroded (and made steeper) by flood water scouring.

Special case of dikes or dike elements made of masonry or concrete

These structures are common in urban areas and are made from masonry stones or reinforced concrete. In most cases their function is also – if not primarily – to support and protect the bank and they are often referred to as "quay walls". They may become unstable as a result of scouring phenomena combined with a design fault (*e.g.* insufficient toe protective works) and/or ageing of the components (*e.g.* degradation of the wall joints or rotting wood piles in the foundation). If these structures fail, they do so abruptly.

Efficient tools adapted to diagnosing these structures and their specific features are currently being developed. One interesting example is the detailed diagnostic study carried out in 2002-2003 on the quay walls of the Rhone river where it crosses through the town of Arles (in the south of France), a project managed by the local dyke owner, the SYMADREM.

Comment: it is important to note that the various degradation mechanisms described above (§ 2.2.1 to 2.2.4.) may act simultaneously or sequentially in a process culminating in the failure of the dike. For example, erosion of the river-side slope increases the likelihood of internal erosion in the backfill if the transverse section of the dike has been significantly worn away or may lead to sliding on the river-side slope (since the overall slope is steeper) when water levels drop after a flood peak.

2.3 Diagnosing a structure

2.3.1 Definition of the diagnosis

By definition, diagnosing a civil engineering structure involves:

– determining how safe the structure is and identifying its weakness, faults and anomalies;

– defining judiciously the work that will overcome or remediate the observed shortcomings.

2.3.2 Particularities of dike diagnosis

The watercourses in France are retained by approximately 8 000 km of dikes. The construction drawings for these very long and winding structures have often been mislaid and in many cases maintenance work has been abandoned. The cost of diagnostic work is therefore very high – due to the length of the structure alone. The difficulties are amplified by the fact that the investigation attempts to identify weaknesses in a structure that is rarely subjected to the hydraulic and mechanical stresses that it is designed to resist.

In common with dams (Degoutte, 1992), dike diagnosis is a two-step process:

– a summary diagnosis that involves clearing, if necessary, bushes covering the dyke and then conducting a topographical survey of the structure and its surroundings followed by a visual inspection. Regular clearing of the vegetation and visual inspections are also essential for the routine surveillance of the dike;

– a detailed diagnosis, including all the phases described in the third part of this guide.

Dividing up the diagnosis into two steps offers two advantages:

– it makes step one more likely to happen or be scheduled, even at a minimum intervention level; this step quickly determines the condition of the structure and the zones at risk (pre-zoning);

– it sets up a framework for the detailed diagnosis, based on the summary diagnosis, that may be limited to the high-risk zones identified during step one.

3

GENERAL METHODOLOGY FOR THE EFFICIENT DIAGNOSIS OF DRY DIKES

3.1 Introduction

There are three elements involved in preventing dikes from failing: **surveillance, maintaining** and **diagnosing**. The first two of these are described in a French guide to flood protection dikes (Mériaux *et al.*, 2004)[3]. Dike diagnosis was covered in a slightly earlier publication (Lino *et al.*, 2000) which introduced the concept of combining geophysical and geotechnical methods to enhance diagnostic analysis. A great deal more has now been learned about these methods – particularly geophysical methods – primarily as a result of experimental work conducted within the framework of the "CriTerre" French National Project (dikes on the Cher and Agly rivers). Within this context, the aim of this guide is to describe these geophysical and geotechnical methods and to assess their ability to provide not only an effective diagnosis but one that can efficiently cover the thousands of kilometres of dikes that run alongside French waterways.

---

3. Soon to be available in English (Mériaux, Royet, 2007).

Dike diagnosis is a three-phase process. The starting point is a **preliminary study phase**, *i.e.* the **search for indicators** that guide the work conducted in the second phase, the **geophysical studies**, and in the third phase, the **geotechnical studies**.

3.2 Preliminary study phase: searching for indicators

As a general point, any investigation of a marine or river civil engineering structure must start with a search for indicators. This phase is a key component of dike diagnosis, and involves collecting as much information as possible about the dike's history, its external characteristics (topography) and its role in the river system (local geology and river dynamics). The key considerations for this preliminary study are a determination of the nature of the component materials of the dike, the nature of the foundation on which it rests and the hydraulic and morphodynamic conditions that it must resist. This phase is an essential prerequisite to the geophysical and geotechnical studies conducted in subsequent phases.

3.2.1 Historical research

Historical research is a key element in establishing a full diagnosis of a dike. A good illustration of its importance is provided by Halbecq (1996) who conducted a detailed analysis of archive data covering 400 km of levees in the middle reaches of the Loire between the confluence of the Allier and Montjean. The study considered the three floods in the 19th century (1846, 1856 and 1866), listed the dike failures and proposed an analysis of the causes of the failures.

Halbecq concluded that:

– about half of the breaches associated with these three floods were caused by over-topping phenomena, and that in about one third of cases the site of the repaired breach was breached again at a later date;

– the morphology (orientation and dimensions) of the "dike-river" system merits particular attention. Indeed, in 30% of cases, hydraulic phenomena resulting from the dike lying perpendicular to the main flow, or the hollowing-out of the bank in a meander, were enough to cause a breach. Moreover, insufficient inter-levee width and the levee being built directly alongside the river bed contributed to the failures.

This study clearly illustrates just how much information historical research can yield. Unfortunately, the time allocated to this phase is all too often limited. As much as possible must be done to investigate the following four key points:

– **find out what data is available:** start with the operator of the dike, and then consult national public organisations that specialise in the study of hydraulic structures and finally universities, civil engineers and consultants who may have already worked on the levees. Additional information may be obtained from an extensive interview of the local population. Any other relevant documents (old maps, national survey maps, topographic drawings and aerial photos) must be investigated;

– locating historical breaches: local and regional authorities generally keep archive information about floods and the damage they have caused. On one of the "CriTerre" Project experimental sites (the levee on the right bank of the Cher river at Savonnières), the position of the historical breach of 1856, mentioned in the archives and unmistakably located by a small lake on the land side, was confirmed by the results of electromagnetic measurements made in 1999;

– locating and recording physical evidence (e.g. high water marks) of historical floods;

– searching for areas used as a source of materials both for dike construction and breach filling. There are two reasons for conducting this search. Firstly, it indicates the nature of the component materials of the dike. Secondly, in the past, materials were often extracted from the immediate vicinity of the dike, thereby creating new areas of weakness whose position needs to be known.

3.2.2 Geological study

This study identifies the materials found locally that are often used to build the dike: most older dikes were constructed using locally-extracted materials (cut and fill earthworks). The geological study must also determine the nature of the foundation on which the dike has been built: the bedrock (*e.g.* karst), especially if it is near the surface, may have a large impact on the dike and its environment. The main aim of the study is to describe the arrangement and characteristics of the geological formations. The nature of the land in many countries has been surveyed in detail, and records are kept by official organisations. For example, in France, this data is regularly updated, and published both in the form of 1:50 000 scale geological maps and as a subsoil database.

For civil engineering purposes, traditional geological data is not sufficient for a number of reasons. On the one hand, the zone being studied, which is generally the near surface, may differ locally from the overall geological zone described in the map. Moreover, changes to the land, either from natural (such as surface run-off) or anthropological processes (such as quarrying), may have significantly modified the structure of the soil. A geological study is thus a necessary prerequisite to any further studies and work.

3.2.3 Morphodynamic analysis

The morphology of the "dike-waterway" system changes with time. Levee zones that were previously unthreatened may be exposed to new risks generated by a change in river profile. For example, the appearance of a sandy islet in the bed of the river will shift the currents to the banks. The purpose of the morphodynamic analysis is therefore to identify and characterise past and future changes to the watercourse channel. To this end, the following points must be fully investigated (Lino *et al.*, 2000):

– for meandering rivers: translation of meanders and the possibilities of meanders being naturally cut-off;

– for braided rivers (primarily): sideways progression of the river arms or deposits;

– for all rivers (meandering, braided, torrential): deepening of the bed by regressive or progressive erosion, hard points or abrupt changes of slope in the longitudinal bed profile.

Sequences of maps or aerial photos that cover a long period of time are very useful for a morphodynamic study.

The zone studied is a belt running for several kilometres and containing the dike, the dike-limited high-water channel and zones on the land and river side of the dike. Theoretical analysis requires an understanding of the sedimentology, hydrology and morphometric characteristics of the waterway.

3.2.4 Topography – Position-fixing data (Lino *et al.*, 2000)

The topography is studied to:

– establish the longitudinal profile of the crest of the dike and to compare it with the river surface profile of flood water levels, so as to assess the risks of overtopping during a flood;

– map the transverse profiles of the dike ahead of the geotechnical studies (stability, risks of piping), and to locate particular structures in plan and elevation (*e.g.* gates, slipways, access points, crest wall gates, inlet of traversing works, etc.);

– provide a tool for dike management and maintenance: assistance with visual inspection, a positioning reference system for the geotechnical and geophysical surveys in plan and depth. This tool is also useful for dike monitoring purposes.

a) Longitudinal profile

The height of the dike is fixed partly with a view to avoiding overtopping (the major cause of failure). This level is based on the basic-stage flood level – an essential piece of data used in sizing these structures. The dike height and the basic-stage level, as well as the kilometre markers (KM) must all be defined using the same positioning reference system: a national height survey (if it exists) for the elevation and a dike KM reference system (pre-existing or created for the study) for the linear positioning. If necessary, the road or river KM system can be used.

The longitudinal profile is determined on the top platform of the dike with an increment length of 20 to 25 m, or less if the crest is irregular. It is very important to identify any low points, which are often the cause of failure due to overtopping.

The top of the freeboard feature, if there is one, may also be profiled to determine the freeboard available compared with the basic-stage flood and to reveal the sections where this freeboard is insufficient.

b) Transverse section

Knowing information about cross sections of the dike is essential for the correct characterisation of the structure and for designing any strengthening. It is also important when assessing the potential instability or failure during a flood, when the hydraulic head exerted on the dike may be as high as 4 or 5 m or more. There are three phenomena associated with this type of failure:

– internal erosion of the dike or of its foundation (piping);

– instability of the land-side slope during the flood;

– instability of the river-side slope when the water level drops.

Transverse sections are investigated every 100 to 200 m for homogeneous zones, and every 50 to 100 m for complex zones. These sections must comprise 8 to 12 survey points as a minimum – more if particular structures are incorporated into the dike – and must extend about ten metres beyond the dike on both the river and land side.

c) TOPOGRAPHICAL MAP

Drawing up a topographical map to a scale of 1:500 or 1:1000 is a critical element in ensuring the quality of the diagnosis and is the basic tool used for monitoring and maintaining dikes. This drawing will be the canvas onto which all the information and observations collected about the dike during the visual inspection and during the geophysical and geotechnical studies will be accurately copied.

This map is usually correlated to a standard reference system (*e.g.* the Lambert grid in France). This simplifies the integration of the results into geographical information systems (GIS) and into other data processing software.

3.2.5 Visual inspection (Lino *et al.*, 2000)

The visual inspection identifies and logs all the external morphological features of the dike in addition to any clues that reveal or suggest the presence of an anomaly: earth movements, erosion zones, gully erosion, a particular type of vegetation, animal burrows, pipes and conduits, etc. The inspector must walk along the entire portion of the dike under investigation, and must cover each part (crest, land- and river-side slopes and dike feet). If the levee runs alongside the river, an inspection from a boat and an underwater inspection may be necessary.

This phase should not be carried out until after the historical research and a detailed topographic survey have been completed. This groundwork will define which type of method should be used for the visual inspection: a kilometre-marker system, markings on the ground, length of the described elemental sections, etc. The visual inspection will confirm or invalidate any information collected previously, and will accurately locate all features of interest on a map.

Moreover, the visual inspection must be carried out under optimal observation conditions. For example, periods of vegetative growth should be avoided if possible. Regardless of the time of year, the undergrowth should be cleared in advance from all zones due to be inspected.

In terms of resources, three persons skilled in civil/geotechnical engineering should carry out the visual inspection.

a) FEATURES TO INSPECT AND INFORMATION TO LOG

The primary purpose of the visual inspection is to verify and complete the information collected during the historical research stage. Additionally, water levels are recorded on either side of the levee on the day of the inspection. Finally, if the dike is fitted with monitoring equipment (piezometers), the measurements should be recorded.

b) INDICATORS OF ANOMALY

– land-side slope
• vegetation (type, spread, roots and stumps),
• signs of soil creep and gully erosion,
• burrows (size, density),
• pipe outflows and one-off structures,
• indicators of seepage, damp areas and water holes,
• existence, type and condition of land-side reinforcement and/or of a protective facing,
• unusual topographical features beyond the foot of the slope (indicators of old breaches, hollows, sand boils, sink holes, trenches, channels);

– crest
• vegetation (type, spread, roots and stumps),
• type and suitability for vehicles of the crest road or path,
• longitudinal and transverse cracks of fissures,
• soil settling, sink holes,
• burrows (size, density),
• existence, type and condition of the freeboard feature,
• existence, type and condition of the facing,
• particular structures;

– river-side slope
• vegetation (type, spread, roots and stumps),
• signs of soil creep, gully erosion, sink holes,
• burrows (size, density),
• pipe outflows and particular structures,
• damp areas, water holes and high water marks,
• existence, type and condition of protective facing,
• existence, type and condition of the protection of the toe of the slope,
• unusual topographical features beyond the toe of the slope (water-level erosion, alluvial deposits, sink holes).

Moreover, constructions (inspection holes, pumping stations, houses, etc.) nearby or incorporated into the body of the dike may constitute weak points that must be identified and located accurately on the study map. At this point, the extensive interview of the local population is carried out if it has not been conducted before.

All the information collected during the visual inspection is copied onto the topographical maps (with a scale of 1:500 or 1:1000). All observed anomalies are marked on the map and numbered. These numbers refer to successive entries in the anomalies record forms where they are detailed and commented on. The transverse sections are generally drawn out on the back of the forms at the corresponding kilometre marker (KM).

A portfolio of photographs is usually compiled to accompany the forms. This portfolio contains one or more views of the entire structure and photographs of each anomaly, referenced to the corresponding anomaly number.

3.3 Geophysical exploration

3.3.1 Principle and purpose of the geophysical exploration

A geophysical exploration of a dike involves deducing the internal characteristics of the structure by studying the variations in a physical field measured along or across the dike (on the crest or at the toe) (figure 12).

The findings provide a picture of the inside of the structure, along its entire length, that reveals the nature and distribution of materials, zones that have been reworked and the presence of heterogeneities, conduits, buried pipes and cables, etc.

Based on an analysis of the results, the structure is divided up into zones: conforming zones and heterogeneous portions that might degrade when the dike is stressed. The assumptions made from the geophysical exploration must be correlated:

– firstly, with the results of the preliminary studies phase;

– subsequently, with the results of soundings (*e.g.* boreholes, ...), the positioning of which may have been chosen based on the results of the geophysical measurements with a view to investigating areas where the constitution of the body of the structure was found to vary.

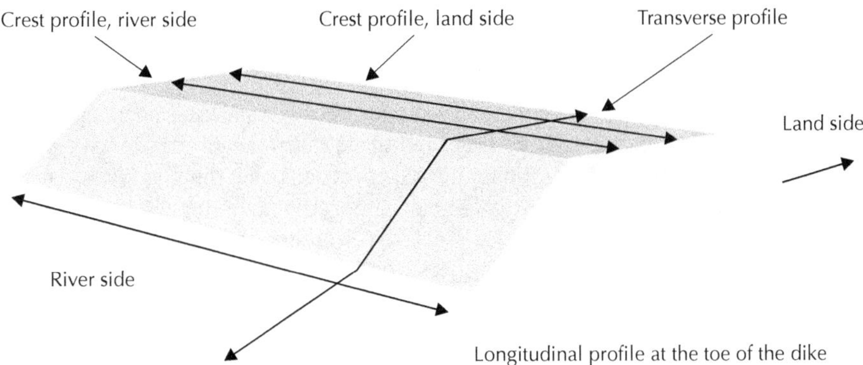

Figure 12 – Profiling paths on a dike

3.3.2 Which physical quantities should be measured for dike diagnosis?

Dikes are very long structures. For the first phase of the exploration, the need to diagnose along lengths of several kilometres demands the use of **efficient methods** that provide information about the characteristics **of the entire body of the structure** down to its foundations. In the context of this guide, an "efficient" method is one that acquires a large amount of good-quality information over a long distance in a short period of time.

In order to explore the entire structure, the depth of investigation must be about 8 to 12 m in France. This covers the average height of a dike plus some of its foundation. Moreover, the materials used to build dikes generally have a high content of loam or clay. These materials conduct electricity, *i.e.* their constituents allow the flow of a direct current or the transmission of low frequency electromagnetic waves. The physical quantity used to measure these phenomena is the conductivity σ of the materials (measured in $S.m^{-1}$). The inverse of the conductivity is more often studied: the **resistivity ρ (measured in $\Omega.m$)**. Resistivity is the most useful quantity to investigate for dike diagnosis, and is measured using **low-frequency electromagnetic methods and geoelectrical methods**.

The resistivity of a material depends primarily on electrolytic conduction phenomena and, to a lesser extent, on electronic conduction. Resistivity values vary greatly depending on the nature of the material. The resistivity of a material also changes as a function of water content, porosity of the propagation medium, how the voids interconnect (the tortuosity) and water salinity. Archie's law is an empirical relationship that summarises the effects of these properties:

$$\rho = a\rho_e \Phi^{-m} S^{-n}$$

where ρ and ρ_e are the resistivities of the material and of the water contained in the pores, Φ is the porosity, S the fraction (by volume) of the pores that contain water, and a, m, and n are constants such that: $0.5 \leq a \leq 2.5$, $1.3 \leq m \leq 2.5$ and $n \approx 2$.

Approximate values of the resistivity (and inversely the conductivity) of the main components of soils are shown in figure 13 and table 1.

Other physical quantities may be analysed:

– the **complex permittivity** of the soil relates to both conduction phenomena and polarisation phenomena. It can be measured using radar or high-frequency electromagnetic methods (> 100 MHz). Under normal conditions only the first two metres of the body of the dike can be investigated by these methods, (although depths of 5 m can be reached in some cases), since radar waves do not propagate easily through the conducting media that form the majority of French dikes;

– the **mechanical impedance** of the materials, measured using seismic methods, identifies reflective boundaries in the structure, and particularly the interface between the loose foundation and the *substratum*. Information about these quantities can be used as a complement to resistivity measurements.

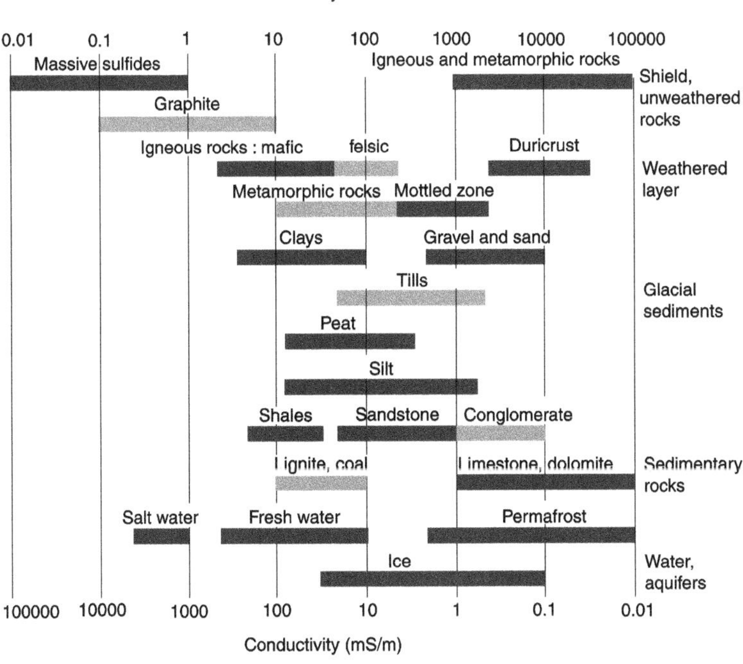

Figure 13 – Range of electrical resistivities of the main earth materials (Palacky, 1991)

Material	Resistivity (Ω.m)	Material	Resistivity (Ω.m)
Meadow, pasture land	10-50	Marly or argillaceous sandstone	15-120
Field, orchard	50-100	Fissured sandstone	1,000-1,500
Fallow land, forest	100-1,000	Compact sandstone	2,000-5,000
Wet gravel	1,000-5,000	Marls	0.5-60
Dry sand	500-10,000	Soft clay, mud	0.5-30
Wet sand	100-350	Clays	1-100
Dry loam	400-2,000	Compact schist	2,000-3,000
Wet loam	20-200	Weathered schist	30-200
Peat	20-700		
Chalk	100-300	Fresh water	100-10,000
Pure limestone	300-5,000	River water	20-60
Cement rock	80-300	Sea water	0.1-0.5

Table 1 – Approximate values of the resistivities of a few materials (Hollier-Larousse, 2000)

3.3.3 Geophysical methods of choice for dike diagnosis

a) Efficient exploration with optimum penetration depth
for dikes whose crest is suited to road vehicles

Low-frequency near-field and far-field electromagnetic methods operate at the optimal depth of penetration for dikes and are the efficient geophysical methods of choice for diagnosing dry dikes on which a vehicle can be driven.

For **near-field methods** the source (an electric coil) is positioned at a fixed distance from the receiver (another coil). They are known as dual-coil methods, or more commonly **Slingram methods.** They measure the apparent conductivity (the inverse of the apparent resistivity) of the materials that form the body of the dike and its superficial foundation. These methods can reveal heterogeneities within a narrow range of resistivity, and are more sensitive to conducting materials than resistant materials.

Far-field methods use existing civil and military transmitters that broadcast in the VLF band (Very Low Frequency (3 kHz-30 kHz)), the LF band, and the MF band that extends up to about 1 MHz. These transmitters are generally located several dozen to several hundred kilometres from the measuring point. They can reveal heterogeneities over a broad range of resistivity and are sensitive to both conducting and resistant materials. Within the context of the "CriTerre" French National Project, the method tested was the **Radio magnetotelluric (RMT) method.**

Two measuring devices, one for each method, were tested on dikes. They are both suited to acquiring measurements efficiently over long distances and their depth of penetration is ideal for dike investigation, *i.e.* they can record the resistivity of the materials from the crest down to the top of the foundation of the dike.

b) Exploration over long distances, with optimum penetration depth,
and lower efficiency for dikes whose crest is unsuited to road vehicles

When the crest of the structure is unsuited to road vehicles, the efficiency is inevitably reduced, with the rate of ground coverage limited to the walking pace of the operator. Low-frequency far-field electromagnetic methods can still be used (Bosch 2001), although they were not tested in walking mode in the context of the "CriTerre" Project.

Slingram methods were, however, found to operate very satisfactorily along the crest and toe of the dike.

c) Local investigation with optimum penetration depth

Once the investigation of the entire length of the structure is complete (a few kilometres or tens of kilometres long), the first step is to zone the dike into successive sections based on the results of the high-efficiency methods (*e.g.* sections with constant value or, on the contrary, with large variations in the geophysical quantity measured). This process identifies several portions (a few metres or tens of metres long) within which geophysical measurements should be performed with greater precision. The final zoning can also be delayed pending the results of specific geotechnical investigations. **The use of a local geophysical method** (*i.e.* a method whose efficiency is low but whose results at a specific location are more accurate than high-efficiency methods) has become **an essential step in the diagnosis.**

High and medium-efficiency methods can, nonetheless, be applied to local investigations so long as the measuring-point increment is adjusted (usually to about one metre) to match the nature of the zone being studied. However, low-frequency electromagnetic methods are restricted to locating heterogeneities at depth and to the accurate imaging of the distribution of the resistivity within the structure.

Geoelectrical methods (2D electrical imaging surveying) are widely used to study the geometry of the dike, its stratification and the distribution of its heterogeneities. The principle is simple: a direct current is injected into the ground via two electrodes and the potential difference is measured at two other electrodes. These methods are more time-consuming, but provide a wealth of information, and they can easily sound the full height of the dike. **They can be used as a follow-up to high-efficiency methods, or alternatively after geotechnical tests at the same location, to fully investigate any correlations between geophysical and geotechnical data.**

Seismic refraction is useful in determining reflective boundaries and heterogeneous zones. Although expensive in dike investigation applications, they are valuable in exploring the contact between the loose foundation of the dike and the *substratum below.*

d) Less effective exploration: other methods

Ground penetrating radar (GPR) offers high efficiency over long profile distances. It can be used on dikes; however, the depth of penetration is poor in conducting media and only allows a small part of the body of the dike to be investigated, since dikes consist mainly of conducting materials. The investigation depth rarely exceeds 2 m. However, for a low dike with a suitable composition, it is an inexpensive, high-efficiency method that provides a wealth of information.

High-efficiency electrical methods (using capacitive electrodes) have been proposed and trialled on dikes suitable for vehicles, but their investigation depth is limited (about 2 m). Other devices of this type may merit investigation.

High-resolution seismic reflection methods and surface-wave seismic methods, fitted with gimble-type sensors, have been suggested. Their efficiency compared with that of low-frequency electromagnetic methods remains to be investigated and tests need to ascertain what additional information this method could provide.

Infrared methods reveal contrasts in surface temperatures. They can be used to detect zones of seepage or upwelling on the land side, or zones that preferentially feed into a drainage ditch.

Spontaneous polarisation methods are also used to detect flows of water through a loaded structure. These methods are not described below since this guide focuses on methods applicable to dry dikes. They should, however, be included in a methodological study designed to diagnose dikes during a flood or in a *post*-flood situation.

3.4 Geotechnical exploration

3.4.1 Principle and objective of geotechnical exploration

Geotechnical exploration produces very accurate information about the component materials of a structure and their distribution. The soil is characterised by tests carried out either *in situ* or in the laboratory on samples taken from the structure. Since the speed of data capture is low, the interval between test points must be adapted to suit the specific nature of each successive section of dike, as determined by the results of the geophysical study. **The principal techniques are: destructive drilling, core drilling, penetrometric tests, permeability tests, shear tests and investigating with a mechanical shovel.**

The objective of a programme of geotechnical exploration is to characterise, at a number of points along or across the dike, defects in the permeability and/or compaction and in the contacts between the various layers. This data must also be used to calibrate the results of the geophysical studies. Consequently, they enhance the interpretation of the collected data and refine the geophysical and geotechnical picture of the structure. They may also identify the need for further local geophysical exploration, such as 2D electrical imaging or seismic refraction.

3.4.2 Geotechnical methods of choice for dike diagnosis

Penetrometric tests are quick and easy to perform. They provide information immediately about the position of the interfaces between layers, and give a qualitative approximation of the mechanical properties of the soils.

Investigating with a mechanical shovel reveals the nature and sequencing of layers of materials in a vertical section (of variable height, but for depths less than 5 m) through the body of the dike. Samples of reworked soil can be taken from the hole or the trench for characterisation in the soil mechanics laboratory.

Permeability tests are more time-consuming and require cumbersome test equipment; however, the information provided by these tests is extremely valuable for dike diagnosis. Modern equipment records additional parameters during the test, such as the speed of advance of the drill bit in the ground, from which certain mechanical properties can be deduced (particularly soil density).

Destructive drilling extracts materials from the dike for visual inspection of their nature. This technique also obtains reworked samples for subsequent testing in a soil mechanics laboratory. **Core drilling** produces intact samples (cores) directly from the structure, which are then prepared for more in-depth analysis in the laboratory. Once again, almost all drilling rigs now record additional parameters to provide real-time information about specific mechanical properties of the soils.

Testing with a phicometer is one of the few tests (together with the vane shear tester) that measure the shear strength of materials in a borehole. It can also measure the shear strength of materials that cannot be sampled by core drilling.

3.4.3 Systematic and/or optimised programme of exploration

A typical programme of geotechnical exploration was described in the context of the methodological guide to the diagnosis of the levees along the middle reaches of the Loire (Lino *et al.*, 2000):

– every 200 m: along the crest of the levee, **penetrometric testing** (static, standard dynamic or light dynamic) down to the foundation (generally 8 to 12 m below the surface). This 200-m interval was selected based on the average size of the breaches created during the floods of the 19th century;

– every 2 km: along the crest of the dike, on the land side, **core drilling** continued into the foundation (12 to 15 m) and fitted with a piezometer so as:
• either to take two or more **intact samples** (for laboratory testing) or to perform three or more **phicometer** tests, one of which was in the foundation,
• to perform two or more **permeability tests**, one of which was in the foundation;

– at the foot of the slope, on the land side: **destructive drilling with recording of various parameters**, continued into the foundation (5 to 10 m) and fitted with a piezometer;

– at a transverse section: **2D electrical imaging**, or ground penetrating radar if necessary, to correlate the geotechnical information with the data from a transverse geophysical profile.

If the only variable of interest is the permeability of the dike, the systematic exploration can be based on the Perméafor device, which is capable of delivering about four tests a day.

Clearly, this type of programme must be adapted to each individual case. The geophysical studies conducted previously must be analysed so as to minimise the number of geotechnical tests performed.

For example, a Slingram profile of apparent resistivity along a long length of the dike may reveal one section of constant values, indicative of the uniform utilisation of materials along this section of the structure, and another section characterised by large variations in resistivity, indicative of a more chaotic use of materials (resulting possibly from repairing a breach). Working with the geophysicist, the geotechnician may select a larger interval between measuring points for the uniform dike section and focus the geotechnical tests with shorter intervals on the non-uniform sections.

Consequently, geotechnical exploration contributes not only to enhancing the interpretation of the geophysical data, but also to refining the geotechnical model of the dike. It is also valuable in defining specific zones that merit investigation by a local geophysical method, such as 2D electrical imaging surveying.

4

GEOPHYSICAL EXPLORATION METHODS

When diagnosing dry dikes, the geophysical exploration begins with a programme of high-efficiency studies conducted using Slingram-type low-frequency electromagnetic methods. Radio magnetotellurics (RMT) or radar-based methods (ground penetrating radar) may also provide useful results: however, RMT is difficult to perform and implement, and the applicability of radar-based methods is limited. The purpose of all these methods is to identify heterogeneous or weak zones of the dike that might respond adversely to flooding.

In many cases, the information provided by low-frequency electromagnetic methods (in the form of a profile) is of a general nature and does not provide a sufficiently accurate picture of the dike geometry. In this case a local geophysical investigation is required to complete this phase, for example using 2D electrical imaging or possibly seismic refraction.

4.1 High-efficiency exploration using the Slingram method

This is a low-frequency near-field electromagnetic (EM) method. The information obtained – about variations in the apparent resistivity – is represented as a profile that provides a picture of the overall distribution of heterogeneities in the levee. Used widely within the framework of the "CriTerre" French National Project, it is the method of choice for the efficient diagnosis of dry dikes. (NB – in the context of this guide, an "efficient" method is one that acquires a large amount of good quality information over a long distance in a short period of time).

4.1.1 Principle

The low-frequency near-field methods described in this guide use magnetic dipoles (coils) as the transmitter and receiver, arranged either vertically or horizontally. An alternating source generates a primary magnetic field at a given frequency *via* an induction coil. When the primary field encounters a conducting anomaly a much weaker secondary field is induced.

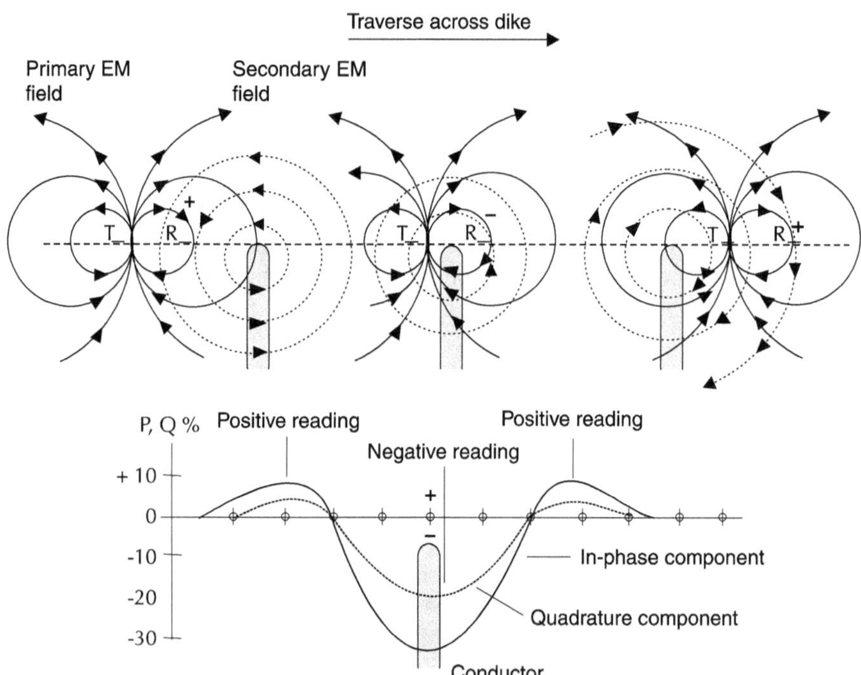

Figure 14 – Principle of low-frequency near-field electromagnetic methods: the setup shown is a Slingram system with a vertical magnetic dipole (2 coplanar coils laid on the ground, HCP configuration or VD mode) (Chouteau, 2001)

4.1.2 Quantity measured by the Slingram method

Measuring the ratio of the quadrature component of the secondary field to the primary field provides a good description of the conductive heterogeneities in the subsoil in terms of apparent conductivity σ_a **(S/m)**, the inverse of the apparent resistivity ρ_a **(Ω.m)**.

For the type of device shown in figure 14 (HCP – horizontal coplanar coils), the quadrature component is directly proportional to the secondary field. At an intercoil spacing (*i.e.* transmitter-receiver separation) that is much less than one wavelength in the ground, this field is 90° out of phase compared with the primary field. These are known as low induction number methods (the propagation factor k at a distance r is such that $|kr| \ll 1$). This characteristic parameter is termed NI and is defined for these methods by:

$$NI = |kr| = s/\delta \text{ where in this case NB} \ll 1$$

where δ (m) is the skin depth defined for the far field method (see the radio magne-totelluric principle) and s (m) the intercoil spacing. The ratio of the quadrature component of the primary magnetic field to the secondary magnetic field is then directly proportional to the apparent conductivity of the ground in the NI formulæ:

$$\frac{|H_s|}{|H_p|} = \frac{NI^2}{2} = \frac{\pi f \mu_0 \sigma_a s^2}{2} \quad \text{and} \quad \sigma_a = \frac{|H_s|}{|H_p|} \frac{2}{\pi f \mu_0 s^2} = \frac{1}{\rho_a} \; (S/m)$$

where $|H_p|$ and $|H_s|$ are the moduli of the primary and secondary magnetic fields respectively, f is the frequency (Hz) of the transmitted wave in the transmitting coil and μ_0 the magnetic permeability of free space ($\mu_0 = 4\pi 10^{-7}$ H/m).

The equations for electromagnetic fields are given by Spies and Frischknecht (1991), for near-field and far-field, in the time and frequency domains, in free space and above a uniform half-space, for the most common Slingram configurations. For more complex devices, software has been developed to model the response in Slingram mode.

4.1.3 Investigation depth

The investigation depth claimed by the manufacturer is the maximum depth at which the device is sensitive in near-field to the presence of a conducting layer. This investigation depth is indicative and strongly depends on the nature of the materials.

This type of device generally operates at frequencies of about a few kHz. The intercoil spacing (varying from a few meters to a few dozen meters) is such that the response to the transmitted field, in the range of apparent resistivity for the most commonly encountered soils, is in near-field. Under these conditions, only the intercoil spacing s and their orientation with respect to the surface have an impact on the investigation depth.

Consequently, for a device fitted with vertical magnetic dipoles **(VD mode), the coplanar coils are horizontal (a configuration also known as Horizontal Coplanar**

– HCP) and the response of the system ($\varnothing_V(z)$) to the presence of a conducting layer located at a depth d is maximal for a depth that is about half the intercoil spacing s ($z = d/S$) and is significant down to **1.5 s (the investigation depth in HCP configuration or VD mode)**.

For a device fitted with horizontal magnetic dipoles **(HD mode), the coplanar coils are vertical (a configuration also known as Vertical Coplanar – VCP),** and it is the near surface layers that make the greatest contribution to the response of the system ($\varnothing_H(z)$) down to **0.5 s (the investigation depth in VCP configuration or HD mode)**.

Consequently, apparent conductivities at depth make a greater contribution to the measurements made in a HCP configuration (VD mode) than in a VCP configuration (HD mode).

Integrating the $\varnothing$ functions as a function of the standardised depth z gives the cumulative responser of the ground (figure 15). These graphs make it easy to interpret the measurements made above stratified media when the number of layers is known (McNeill, 1980a and 1980b).

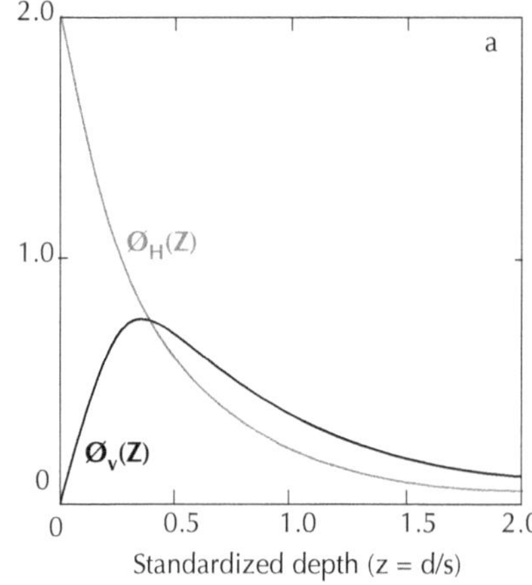

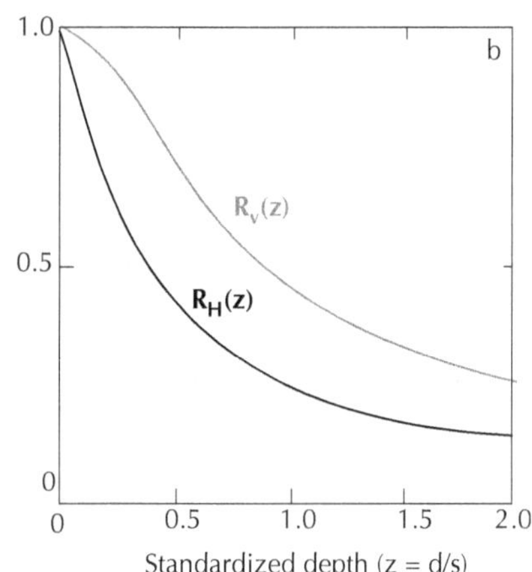

Figure 15 – *a)* Response of vertical and horizontal devices* to a conducting layer at a depth d as a function of the standardised depth z = d/s where s is the intercoil spacing.
b) Integrating these responses over the depth z $R_H(z)$* and $R_V(z)$*, gives the cumulative response for each section of ground for both types of device.
* Without dimension

4.1.4 Output

The output is illustrated in figure 14, representing the ratio of the quadrature component of the secondary field to the primary field. Instead of this ratio, these profiles are considered in terms of apparent conductivity (S/m) or, more commonly, apparent resistivity (Ω.m), whose qualitative representation of information is identical to that provided in figure 16.

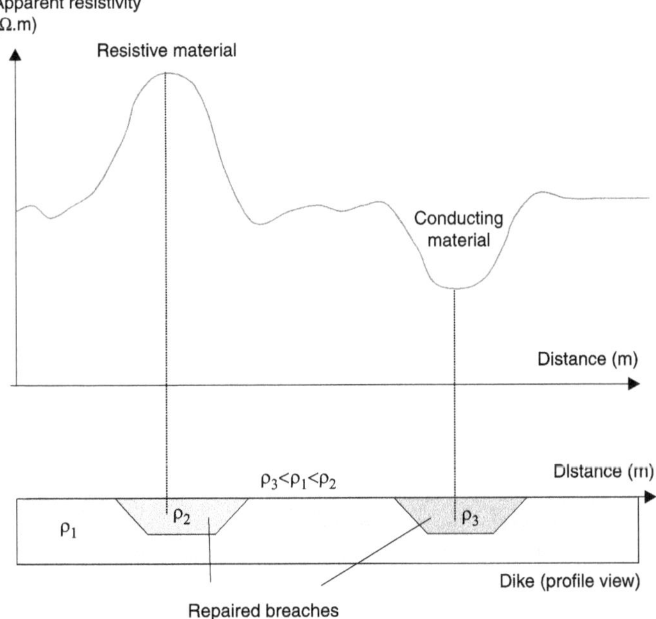

Figure 16 – Qualitative representation of a Slingram profile above a dike that incorporates a resistive and conducting anomaly

4.1.5 Methodology

The intercoil spacing is chosen based on the penetration depth required. Different depths can be investigated by selecting a suitable device (examples for Slingram method types EM31 and EM34 – produced by Geonics – are given in table 2).

	Orientation	Offset (m)	Frequency (Hz)	Penetration depth (m)
EM31	H	3.66	9,800	3.7
EM31	V	3.66	9,800	5.5
EM34	H	10.0	6,400	10
EM34	V	10.0	6,400	15
EM34	H	20.0	1,600	20
EM34	V	20.0	1,600	30
EM34	H	40.0	400	40
EM34	V	40.0	400	60

Table 2 – Parameters for low-frequency dipolar devices EM31 and EM34. The offset is the fixed distance between the transmitting and receiving coils (intercoil spacing).

The coil should ideally be aligned in the direction of the dike when determining both longitudinal and transverse profiles. This avoids interpretation errors caused by the topology of the dike and the nature of its surface materials, since both are generally homogeneous in the longitudinal direction. This means that the boom of the EM 31 device will point in the direction of displacement when taking longitudinal measurements on the crest or foot of the dike.

The profiles are generally determined at the crest and foot of the dike, on the land and river sides, and ideally in the two polarisation modes (VD and HD).

The interval between two measuring points is generally 5 m. This may be reduced (*e.g.* to one metre) if more detail is required. The rate of ground coverage for point-by-point data acquisition when using a device that can be carried by a single operator at a constant height (about 1 m) above the ground varies, but is slightly less than walking speed. This technique is useful for sections that cannot be driven along or that are difficult to access. Larger intercoil spacings are possible, but require two operators. The rate of ground coverage is then reduced and is more difficult to quantify.

An EM31 high-efficiency device was tested in the spring of 2004. The device is small enough to be towed behind a vehicle, but since it is sensitive to metal objects, it was towed about 6 m from the vehicle. The towing speed was about 5 km/h (figure 17) and the device was laid on a drag mat. In other experiments, measurements were made continuously and while walking, with GPS used for measurement positioning.

High efficiency mode: 1 meas/s

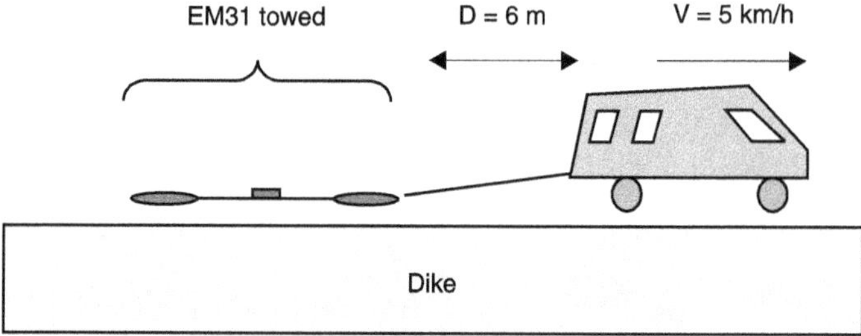

Walking mode: 1 meas/5 m, or continuous with GPS

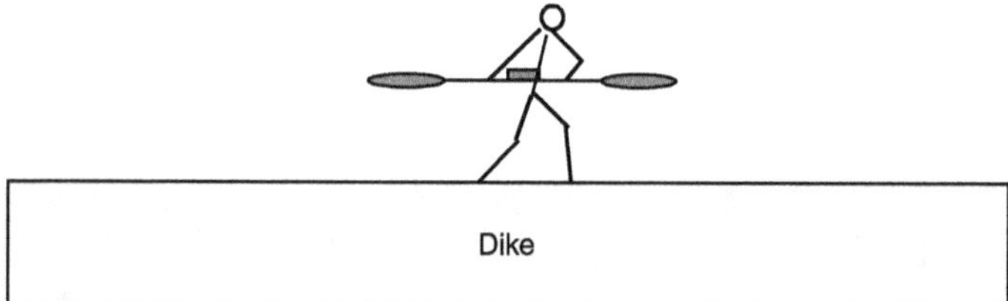

Figure 17 – Measuring modes used on a dike with the EM31 device

The high-efficiency measuring mode tested may be considered for devices that reach greater depths. Experimental development work is currently being conducted in this field.

4.1.6 Interpreting the results

The interpretation is initially qualitative: the presence of a conducting anomaly increases the apparent conductivity measured (and, inversely, reduces the apparent resistivity measured). Moreover, the conducting anomaly induces a local reduction in the ratio of the amplitude of the vertical field to the horizontal field, in both phase and quadrature. A quantitative first approximation may be made using the data collected during the preliminary studies. This approach has clearly revealed **zones filled to repair breaches and damage caused by overtopping**. Moreover, profiling on both the land and river sides of the dike crest may detect transverse anomalies.

These initial measurements are valuable in identifying the locations for geotechnical soundings. The results of these soundings (in terms of the type and depth of the materials) can be extrapolated to the geophysical profiles and can also provide additional quantitative information about the extent and depth of the anomalies. Analysis of the results can be refined through the use of modelling and inversion methods. Generally-speaking, for an equivalent contrast of resistivity between the enclosing material and the anomaly, a conducting anomaly is easier to detect than a resistive anomaly.

4.1.7 Examples of measurements

The measurements presented below were made using a Geonics EM31 device at Saint-Laurent-de-la-Salanque, on the dike on the left bank of the Agly river, a Mediterranean coastal river. Profiles were obtained on the land and river sides, along the side of the road on the crest of the dike, using two measuring methods. The investigation depths were approximately 4 m in HD mode (VCP configuration) and 6 m in VD mode (HCP configuration).

The breached section at Saint-Laurent (at 400 to 600 m, shown in figure 18) was filled with materials of varying coarseness and has a clearly visible signature. Further downstream are other anomalies with higher resistivity in a section where overtopping events are known to have occurred. Overall, the measurements at deeper levels indicate lower resistivity since they include the more conductive materials located further down in the dike body. The roadway, consisting of more resistant bituminous and gravelly materials, provides a greater contribution to the apparent resistivity values for the shallower measurements (HD mode or VCP configuration).

Two data acquisition modes were tested: **walking mode** with one measurement made every 5 m, and **high-efficiency mode** with one measurement every 1 s. The equipment was aligned in the direction of the dike, and towed behind a vehicle at a speed of 5 km/h (*i.e.* with a measurement point every 1.45 m). The experiment lasted for half a day. The point-by-point measurements are compared with the high-efficiency measurements in figure 19. A close-up of the breached section is also shown in figure 19.

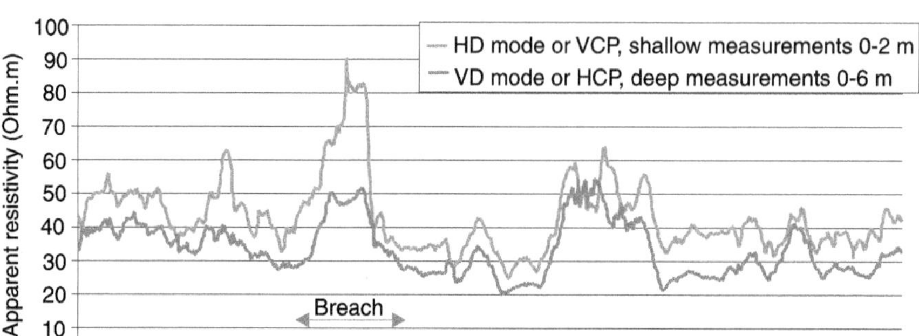

Figure 18 – EM31 apparent resistivity profiles, left bank, land side of the Agly dike

Profiles determined on the land and river sides (figure 20) also reveal the degree of heterogeneity of the dike in the transverse direction. However, the value of these results is limited, since it is not generally possible to determine more than two profiles along the crest of a dike.

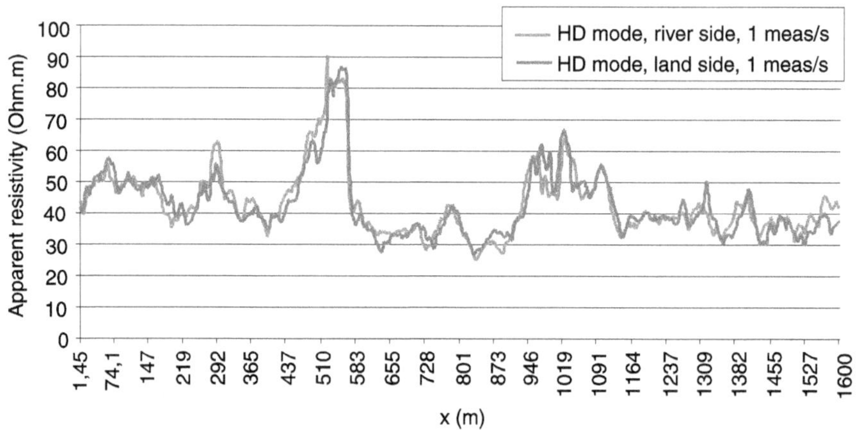

Figure 20 – Two Slingram profiles in HD mode, on the land and river sides, used to assess the degree of transverse heterogeneity of the dike

a) Apparent resisivity
(Ohm.m)

b) Apparent resisivity
(Ohm.m)

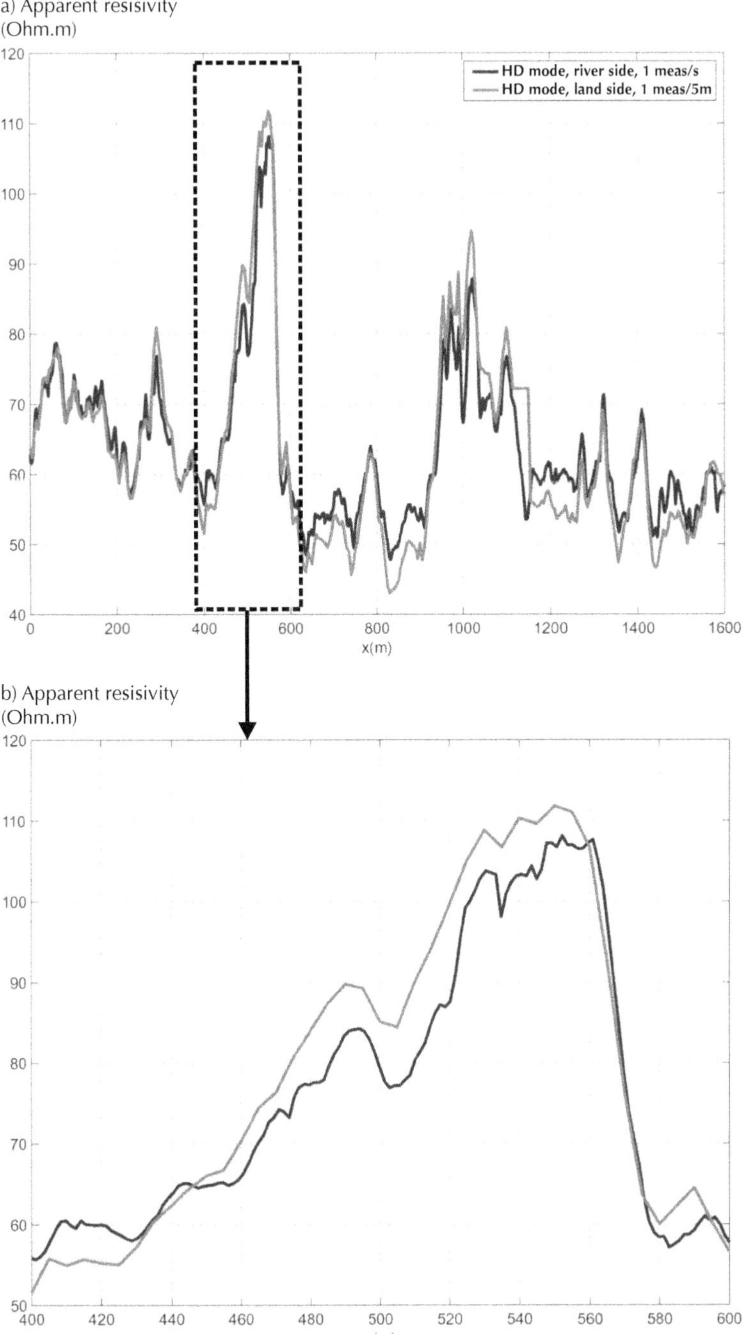

Figure 19 – Comparison of a profile obtained by measuring one point every 5 m, and a continuous profile with one measurement per second

The following example is taken from geophysical measurements made on the dikes on the Cher River (Wakselman, 1999), using a Geonics EM34 device with a measuring increment of 5 m and an intercoil spacing of 10 m. The dike structure is shown in figure 21.

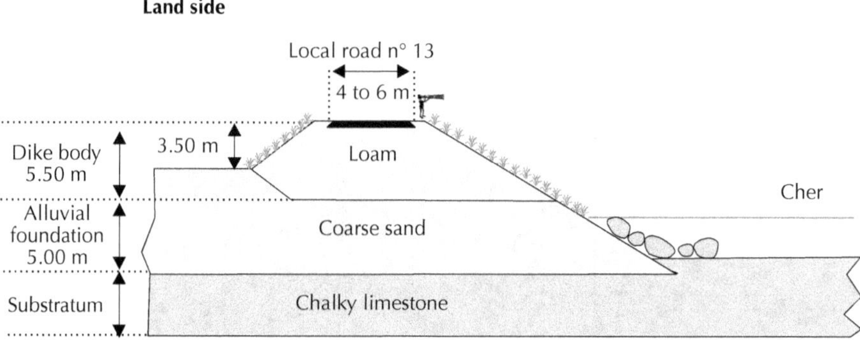

Figure 21 – Transverse section through the dike on the Cher (Grand Moulin hamlet, right bank of the Cher, downstream of Tours)

The measurements (figure 22) made at two different investigation depths (about 7 and 15 m) clearly show that the materials closer to the surface (the loam of the dike body) are in this case less resistant than the deep layers (sandy alluvial foundation and chalky *substratum*). They also reveal the presence of a number of features: a buried conduit, near the surface (3 m deep) and spotted during the visual inspection; a breached section, with a higher resistance and comprising a more conducting anomaly, the cause of which is not clear; and a zone of low resistivity at the end of the profile, corresponding to the presence of a railway ballast connected to the body of the dike: the presence of a conducting element (the railway line) affects the measurements strongly.

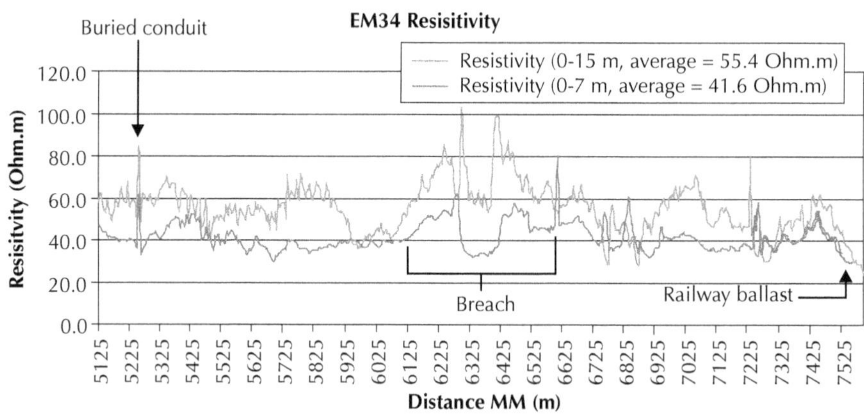

Figure 22 – EM34 measurements at two investigation depths on the dike on the Cher (EDG document)

4.1.8 Conclusion

Low-frequency, dipole-field electromagnetic methods operate at the optimal penetration depth for effective investigation of the body of a dike and the upper portion of its foundation. Furthermore, different investigation depths can be explored by adjusting the coil orientation and the intercoil separation. The measuring interval can be adjusted to suit the degree of detail required, down to about one metre. Their rate of ground coverage in walking mode is slightly less than the walking speed of the operator(s). High-efficiency versions could be used on the crest of a dike if the crest is suited to vehicles, and this possibility has been tested successfully on a dike on the Agly river, for a specific device whose maximum investigation depth was 6 m. The possibilities offered by equipment towed behind a vehicle, exploring depths of at least 10 metres, remain to be studied. It is imperative that the interpretation of the results is correlated with the preliminary studies and is carried out by an experienced geophysicist.

4.2 High-efficiency exploration using the radio magnetotelluric method

Radio magnetotellurics (RMT) is a low-frequency, far-field electromagnetic method.

It uses the military and commercial transmitters erected for radio communications in the 10 kHz to 1 MHz band (the VLF radio band extends from 3 to 30 kHz, the LF band from 30 to 300 kHz, and the MF band from 300 kHz to 3 MHz).

Although RMT has been the most widely used high-efficiency method in the "Cri-Terre" French National Project, difficulties have become apparent relating to the implementation of the method and the interpretation of the data.

4.2.1 Principle

a) PRIMARY AND SECONDARY FIELDS

The radio transmitters used as the source field for the method are considered as far-fields: the electromagnetic waves received at the measuring point, *i.e.* at the crest or foot of the dike, are plane waves. These waves propagate through the surface of the soil. If a conducting or resistant anomaly is present, the induced field detected at the surface is modified. Three phenomena are involved (Chouteau, 2001):

– the production by a source of a primary electromagnetic field, that varies with time;

– this field induces eddy currents in the ground;

– these currents are modified by the presence of heterogeneities.

For a 3D heterogeneity with a distinct outline, two complementary mechanisms may be involved (West and Macnae, 1991; Bourgeois, 2000):

– an **induction effect**, produced by the flux of the primary magnetic field across the heterogeneity (Faraday's law). This phenomenon is termed a **vortex effect** since it generates eddy currents inside the body. Only conducting bodies produce this effect.

– a **static effect**, produced because the heterogeneity deviates the eddy currents induced in the host medium; this effect is termed a **galvanic effect** since the principle underlying its generation is identical to that of electrical methods. The deviation of the currents concentrates the currents towards the inside of the body if it is relatively conductive, or conversely causes the currents to pass around a body that is relatively resistant.

NB: the basic principle underlying plane wave methods is the measurement of one or more components of the total electromagnetic field (the primary one plus the secondary from heterogeneities) due to the galvanic effect (McNeil and Labson, 1991, pp. 576-577). It is only in highly resistive host media (> 10^4 Ω.m) that the vortex effect from heterogeneities could be taken into account. These methods are thus comparable to direct current electrical methods. In this respect, they are sensitive not only to the presence of conductive objects, but also to resistant objects.

b) Polarisation

Finally, the response from an anomaly in the ground depends on the orientation of its major axis compared with the direction of the primary field. There are two types of polarisation:

– **H polarisation,** whereby the primary magnetic field is parallel to the major axis of the heterogeneity. The response of the body is expressed primarily by its static effect;

– **E polarisation,** whereby the primary magnetic field is perpendicular to the major axis of the heterogeneity. The response of the body is expressed as the generation of a secondary electromagnetic field, whose vertical magnetic components are measured, in phase and quadrature with the horizontal component, using coils above the ground.

4.2.2 Quantities measured by RMT

RMT measures (figure 23) the secondary horizontal electric field $\mathbf{E}_x$ induced in the ground, and the horizontal magnetic field $\mathbf{H}_y$ that is the resultant of the primary field $\mathbf{H}_{py}$ and the secondary field $\mathbf{H}_s$ (in general, $\mathbf{H}_s \ll \mathbf{H}_{py}$).

The complex ratio $\mathbf{E}_x/\mathbf{H}_y$, namely the surface electromagnetic impedance, can be used to determine the following variables:

– the **apparent resistivity** (in Ω.m), obtained using the equation formulated by Cagniard (1953), and illustrated along the length of a profile. The apparent resistivity is equal to the resistivity of a homogeneous volume of soil that is equivalent to the sum of the actual values for this volume (*i.e.* taking into consideration measurements of **E** and **H**):

$$\rho_a = \frac{1}{2\pi\mu_0 f} \frac{|\mathbf{E}_x|^2}{|\mathbf{H}_y|^2}$$

where μ_0 is the magnetic free-space permeability, $\mu_0 = 4\pi 10^{-7}$ (H.m^{-1}), and f (Hz) is the radio frequency used.

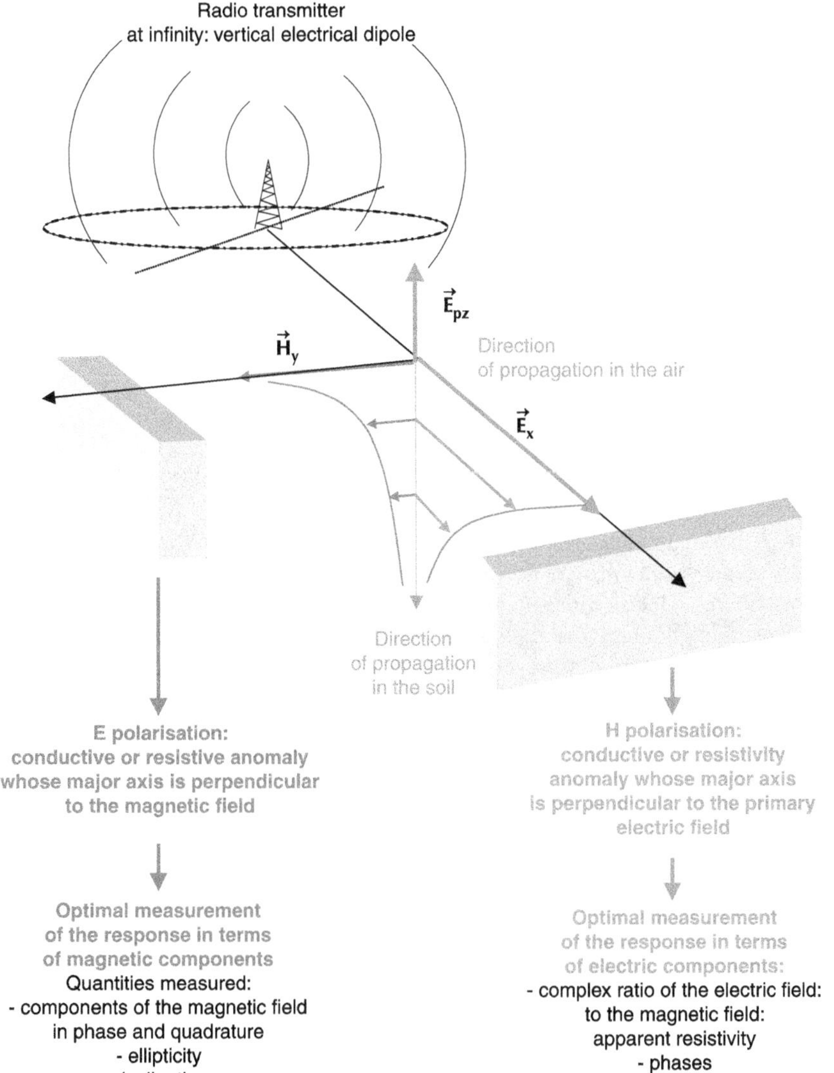

Figure 23 – Principle of low-frequency, far-field electromagnetic methods, polarisations and measuring modes

– the **phase** of the electric field compared with the magnetic field (= phase shift).

When the ground is homogeneous, the apparent resistivity is equal to the actual resistivity. The phase angle φ is 45°. Any other value is indicative of the presence of a heterogeneity: a phase angle of less than 45° indicates a "conductor on resistor" sequence and a phase angle greater than 45° indicates a "resistor on conductor" sequence.

4.2.3 Correcting the measurements in resistivity mode: verticalisation and invariants

The polarisation of the primary field introduces an anisotropy that elongates the apparent resistivity anomaly measured in a direction perpendicular to the primary electric field. There are two ways of correcting this effect:

– when a single transmitter is available, the **verticalisation** of the electric field E_x (Tabbagh *et al.*, 1991) produces resistivity maps that provide a more accurate description of the distribution of the bodies in the ground;

– the use of **two orthogonal transmitters at similar frequencies** is preferable. It has been demonstrated (Guerin *et al.*, 1994) that it is preferable to calculate one or other of the following two invariants:

$$\rho_{inv1} = \left(\frac{\sqrt{\rho_{a1}} + \sqrt{\rho_{a2}}}{2} \right)^2 \quad \text{or} \quad \rho_{inv2} = \sqrt{\rho_{a1}\rho_{a2}}$$

where ρ_{a1} is the apparent resistivity measured with transmitter 1, ρ_{a2} is the apparent resistivity measured with transmitter 2, and where transmitter 1 and transmitter 2 are orthogonal. Unless a multi-sensor and multi-frequency instrument is available, this processing doubles the time taken to perform this method, but does improve the quality and interpretation of the results. Moreover, measurements made with transmitters at 70° and 50° demonstrated that the calculation of the invariants still gave good results.

4.2.4 Penetration depth

An important concept is the penetration depth (or skin depth), the depth at which the amplitude of the fields is reduced to $1/e$ (*i.e.* 37%, since e~2.718) of its surface value; it is expressed by:

$$\delta = \sqrt{\frac{\rho}{\pi\mu_0 f}} \approx 503 \sqrt{\frac{\rho}{f}} \; (m)$$

The **investigation depth** (figure 24) for low-frequency, plane-wave methods is generally considered to be **half the penetration depth**.

At a given frequency, the greater the resistance of the medium, then the greater the penetration depth.

For a given piece of ground, the lower the measuring frequency, then the greater the penetration depth (figure 24).

Example: for a homogeneous dike body with a resistivity of 50 Ω.m, and a frequency of 162 kHz, the skin depth is approximately 9 m, and the investigation depth will be approximately 4.5 m.

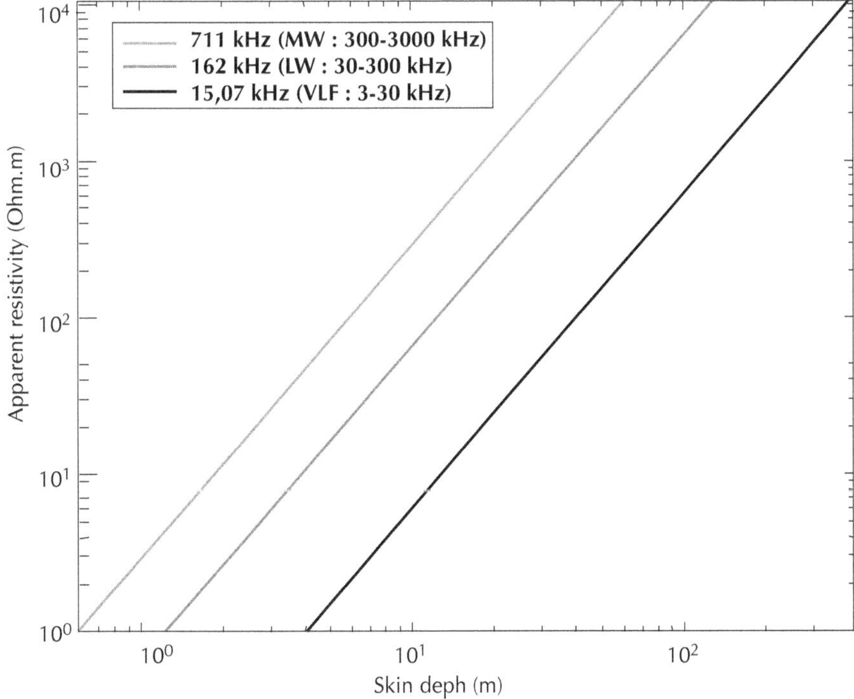

Figure 24 – Penetration depth as a function of the apparent resistivity for some typical frequencies

4.2.5 Output

The output (figure 16) is a profile that shows the apparent resistivity (in Ω.m) of the dike as a function of distance travelled on the surface. The resistivity scale is generally logarithmic.

4.2.6 Methodology

To obtain the most information, measurements should be made in the two polarisation modes (H and E), which requires using two transmitters operating at similar frequencies whose fields are perpendicular to each other at the point of measurement. After estimating the average apparent resistivity of the dike, two other frequencies should be selected so that, in each polarisation mode, there is one investigation that relates to the body of the dike and another that relates to the foundation. For a given frequency, the land and river side should then be profiled, bringing the total number of profiles up to eight or more. If other frequencies are available, then additional profiles can be determined to increase the amount of information collected.

In practice, it is rare to be able to use two orthogonal electric fields at similar frequencies. Usually, only two or three frequencies are available, which makes it difficult to take into consideration resistivity invariants or verticalisation.

Repeatability tests conducted on roadways have demonstrated the stability of the method (Chevassu *et al.*, 1990; figure 25). This is not the case with dikes: experiments performed on the Agly river using various devices have clearly shown that at least three profiles should be produced at a given frequency and along a specified path (Fauchard and Mériaux, 2004; figures 29 and 30). This reduces the efficiency and experience has shown that the method is highly sensitive to surface heterogeneities, which produce peaks of apparent resistivity that introduce "noise" into the measurements and mask the distribution of deeper materials.

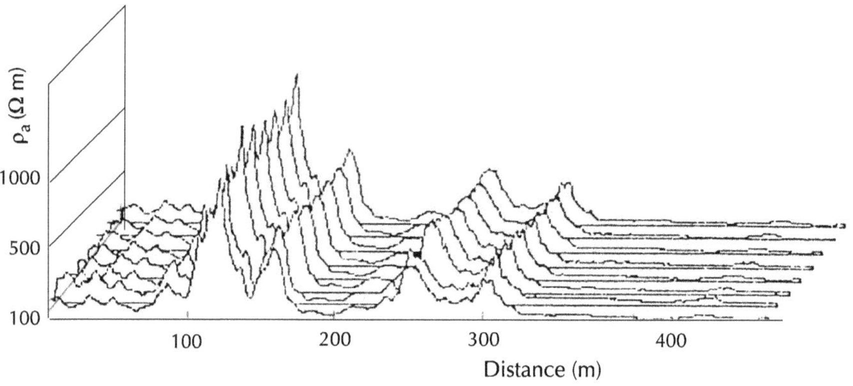

Figure 25 – Repeatability of RMT measurements on a roadway (8 successive runs at 711 kHz)

The measurement device shown in figure 26 is an instrument belonging to the LCPC. Other instruments are also available (Bosch and Müller, 2001).

The electric field is measured using a cable a few metres long, **positioned parallel to the direction of the incident electric field** and connected to a pair of conductive or capacitive electrodes in contact with the ground. For the device described, the contact is capacitive. The magnetic field is measured using a copper coil that tilts automatically in response to the intensity of the magnetic field detected, which ensures that the mat of capacitive electrodes is aligned parallel to the incident electric field. A set of filters and sensors allows the selection of the frequency, and the amplification and the determination of the apparent resistivity.

The mat, coils and electrodes are towed behind a vehicle, thus providing a high-efficiency solution for dikes, whose crest can be driven along. Tests at speeds of up to 30 km/h have been carried out on roads (Chevassu *et al.*, 1990), although the usual speed is about 5 km/h.

At each measurement point, a thumbwheel switch triggers the recording of the electric and magnetic fields on a data logger, usually a PC, connected to the data acquisition chain.

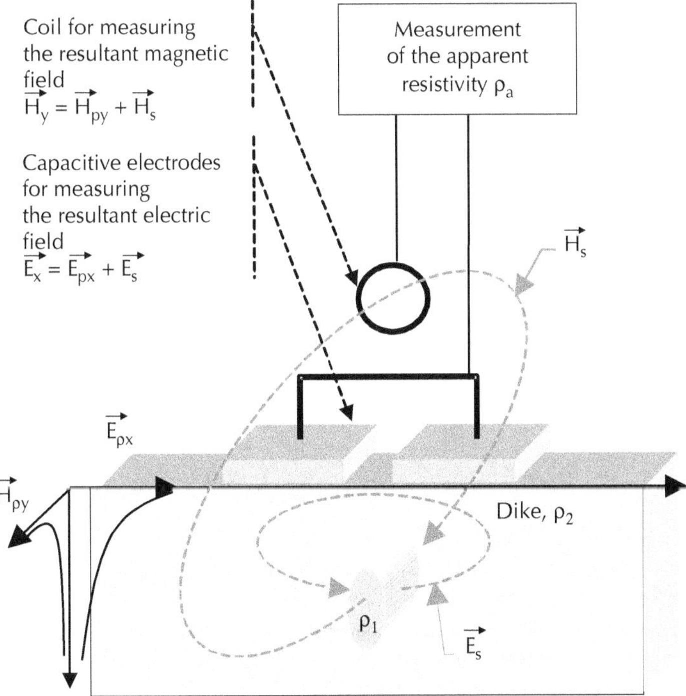

Figure 26 – Principle of the LCPC's RMT device

The separation between the electrodes is chosen depending on the degree of detail required (figure 27). A gap of one metre between the electrodes provides a more detailed investigation of the body of the dike. Profiles determined with a range of mats are shown in figure 31.

For capacitive sensors, the ground coverage must also be considered: the presence of vegetation and of topographic irregularities will affect the quality of the measurements. For this reason, the RMT method in capacitive mode is used if the crest of the dike is suitable for vehicles. If this is not the case, conductive electrodes may be used, along with methods in inclined mode, or VLF-EM methods.

4.2.7 Interpreting the measurements

The initial interpretation of the measurements is **qualitative**. Although difficult, the depth and size of the anomalies found can be characterised, other than in particular cases, such as determining the thickness of a conductive overburden on a resistive layer.

The **quantitative** interpretation begins by considering the results in the light of the results of the preliminary studies. The experiments carried out in the context of the

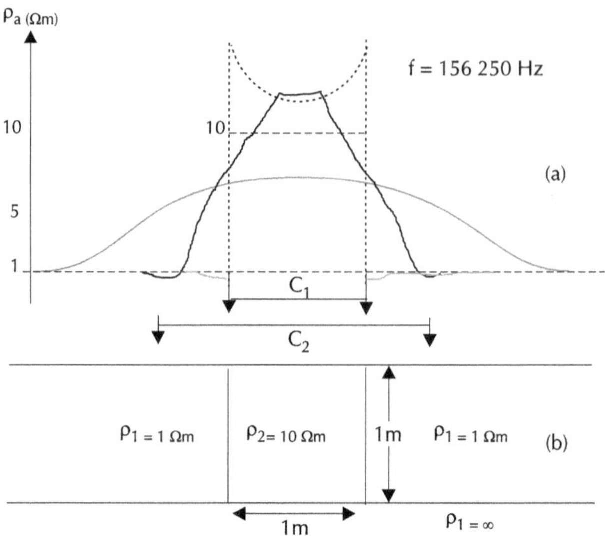

(a) theoretical resistive anomaly (.....) above a two-dimensional structure shown in (b) (H polarisation)

——— anomaly observed with a C$_1$ electric field sensor, 1 m long
········ anomaly observed with a C$_2$ sensor, 2 m long
········ actual resistivity profile of the ground on the surface

Figure 27 – Apparent resistivity anomaly produced by a one-metre wide structure as a function of the distance between sensors (Lagabrielle, 1986)

"CriTerre" Project revealed many correlations between what was found during the visual inspection and the resistivity values measured. Particularly, they highlighted **longitudinal heterogeneous zones** in the dike, such as **breached sections** and **zones where overtopping had occurred**. Indeed, flood-damaged zones had generally been backfilled with brought-in materials. Compared with the original materials, the nature of the newer materials and the construction techniques used were very different – these disparities were picked up immediately by the measurement of the apparent resistivity. Moreover, **transverse heterogeneities such as buried pipes and cables** are also revealed clearly by RMT measurements when at least two profiles are determined on the crest on the river and land sides.

Modelling and **inversion** of the measurements provide additional qualitative information. A prerequisite for this work is some knowledge of the physical structure of the soil, such as the thickness of the overburden or the nature and approximate depth of the anomalies.

However, the effect of the topography of the dike body on the incident field cannot be ignored: the field can reach a heterogeneity in the structure from two directions – directly from the crest and by transmission of the incident field on the slope. So long as the orientation of the dike with respect to the incident field is constant, and so long as the gradient of the slope is constant and uniform with respect to the incident field, then this transmitted field will generate a uniform response at the crest along the length of the structure. On the other hand, any variation in the nature of the materials in the slope, or any variation in the gradient of the slope or a change in the orientation of the dike compared with the incident field (dike sinuosity) will affect the values measured at the crest. **This problem requires further study**.

Conducting a programme of geotechnical soundings has proven to be essential in completing the qualitative interpretation. Analysis of these soundings enhances the interpretation of the profiles in terms of the nature, thickness and mechanical properties of the component materials.

4.2.8 Examples of measurements

The following RMT measurements were made in 1999 and 2004 (Hollier-Larousse, 2001a and 2001b; Fauchard and Mériaux, 2004) on the same dike as that described for the near-field methods: the dike along the Agly river at Saint-Laurent-de-la-Salanque.

Figure 28 (Mériaux *et al.*, 2003) shows two typical profiles determined at 162 kHz on the dike crest, on the land side and on the Agly river side. If the apparent resistivity is considered to be about 50 Ω.m, then the skin depth is about 9 m, and the penetration depth is about 4 to 5 m: at this frequency, it is primarily the dike and the top of its foundation that are explored, since the Agly dikes are between 2 and 3.5 m high.

The average apparent resistivities for each profile are 42 and 52 Ω.m, which reflect the overall homogeneity of the body of the dike. Local anomalies stand out clearly on the land side where **overtopping occurred** and where the slopes of the dike were repaired with coarser materials after the flood of 1999.

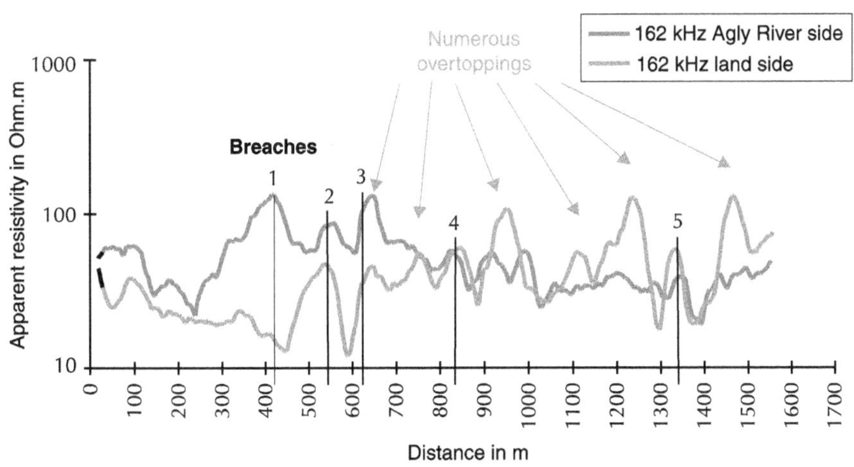

Figure 28 – Example of RMT measurements on the dikes on the Agly River (Mériaux *et al.*, 2003)

The known breach is clearly visible on both profiles. Very different resistivity values (20 and 60 Ω.m), on the land side and on the Agly side, between 250 and 750 m (on either side of the breach), cannot be explained by the visual inspection. They may be a remnant of crest profiling work undertaken after the flood of 1992.

Finally, aligned anomalies on both profiles are indicative of **transverse heterogeneities** (the clearest are indicated by numbered vertical lines). Other than the breach, these anomalies may correspond to trenches and/or conduits crossing the body of the dike or the boundaries of roadway repair work covered by the rolling surface.

The graphs in figure 28 were smoothed with a sliding average covering 40 test points, 1 m apart. Solid experience of the method and a summary of the preliminary studies were required to produce these results.

Raw measurements are more difficult to interpret and at least three profiles, at a given frequency, are essential for each run, as illustrated in figure 29. The breached section, shown in the previous figures, is analysed in more detail here. The four profiles shown, 200 m long, were determined at a speed of 5 km/h, with a measurement interval of approximately 0.5 m. These profiles (figure 29*a*), determined on the same run on the river side, reveal significant inter-profile variability for sections just a few metres long. Moreover, similar variations are shown by all the profiles for distances of tens of metres or more. A smoothing performed with a sliding average of 10 m (figure 29*b*) reveals clearer correlations and offers a more reliable interpretation. Two anomalies at 470 and 590 m are clearly located for each of the four profiles, with an accuracy of a few metres. They form a boundary around a more resistant zone, the result of repair work performed with gravelly materials.

The same measurements and processing were performed on the land side. Two conducting anomalies are clearly apparent on all the profiles; one at 490 m and the other at 550 m. They are both located in the breached section and probably correspond to the presence of pipes or conduits or to the limits of this zone. However, there is a great deal of variability for the first 50 metres of the profiles (figure 30).

4.2.9 Conclusion

The RMT method surveys at the optimal penetration depth since the frequencies generally available allow exploration of the dike and some of its foundation, with a measurement interval of about one metre. Its efficiency can be very high if the device is fitted with capacitive sensors and if it can be towed behind a vehicle along the crest.

The interpretation of the raw results is initially qualitative and identifies the boundaries of zones of different resistivity, on a scale of tens of metres. Determining several profiles along the same path does, however, reveal a broad variability for sections of just a few metres. It is therefore advisable to determine at least three profiles for any one path; this may reveal strong correlations for one-off anomalies that can be located to the nearest metre, and can confirm the general trends of resistivity variation over lengths of tens of metres. This variability originates from two sources:

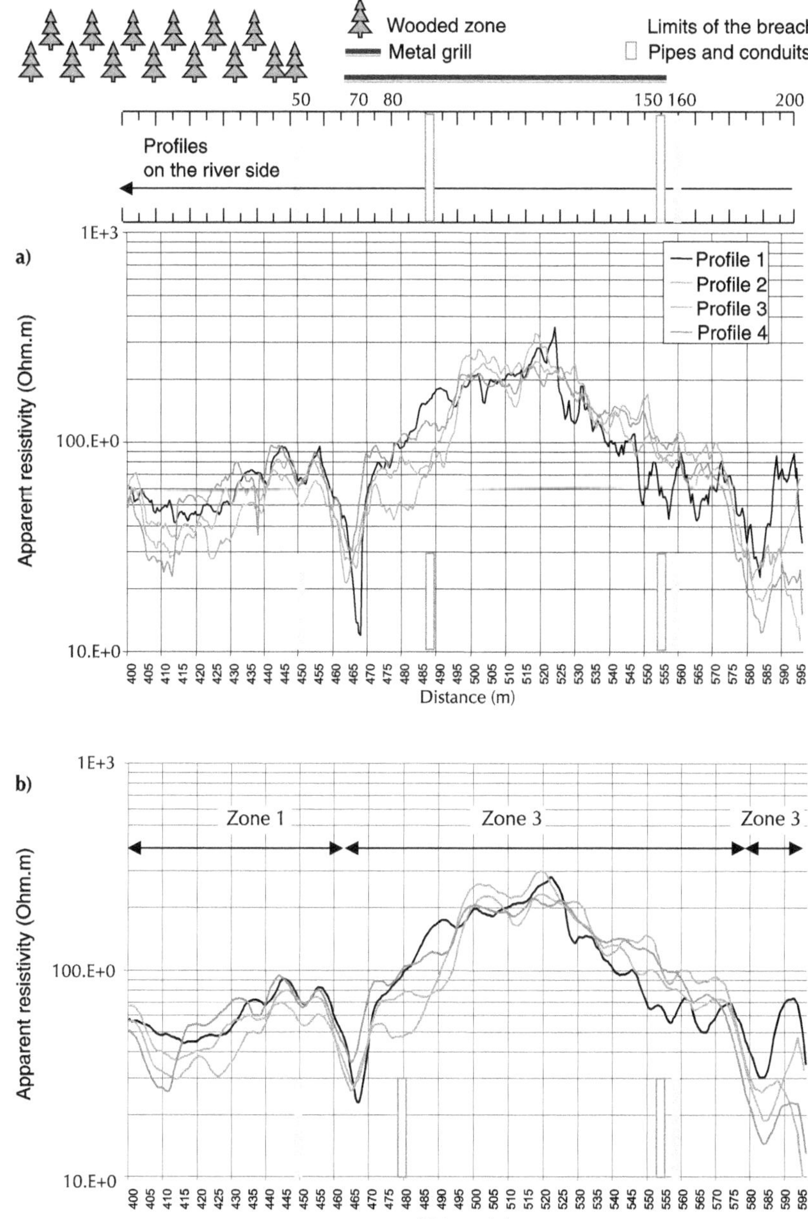

Figure 29 – Comparison of four RMT profiles on the river side. *a)* raw profiles, *b)* filtered profiles

firstly, the high sensitivity of the method to surface heterogeneities and, secondly, topographical effects for which there is currently no solution. Finally, the benefits of using capacitive electrodes is the subject of debate. Experience has shown that they have a tendency to "charge up" over time and that they are sensitive to variations

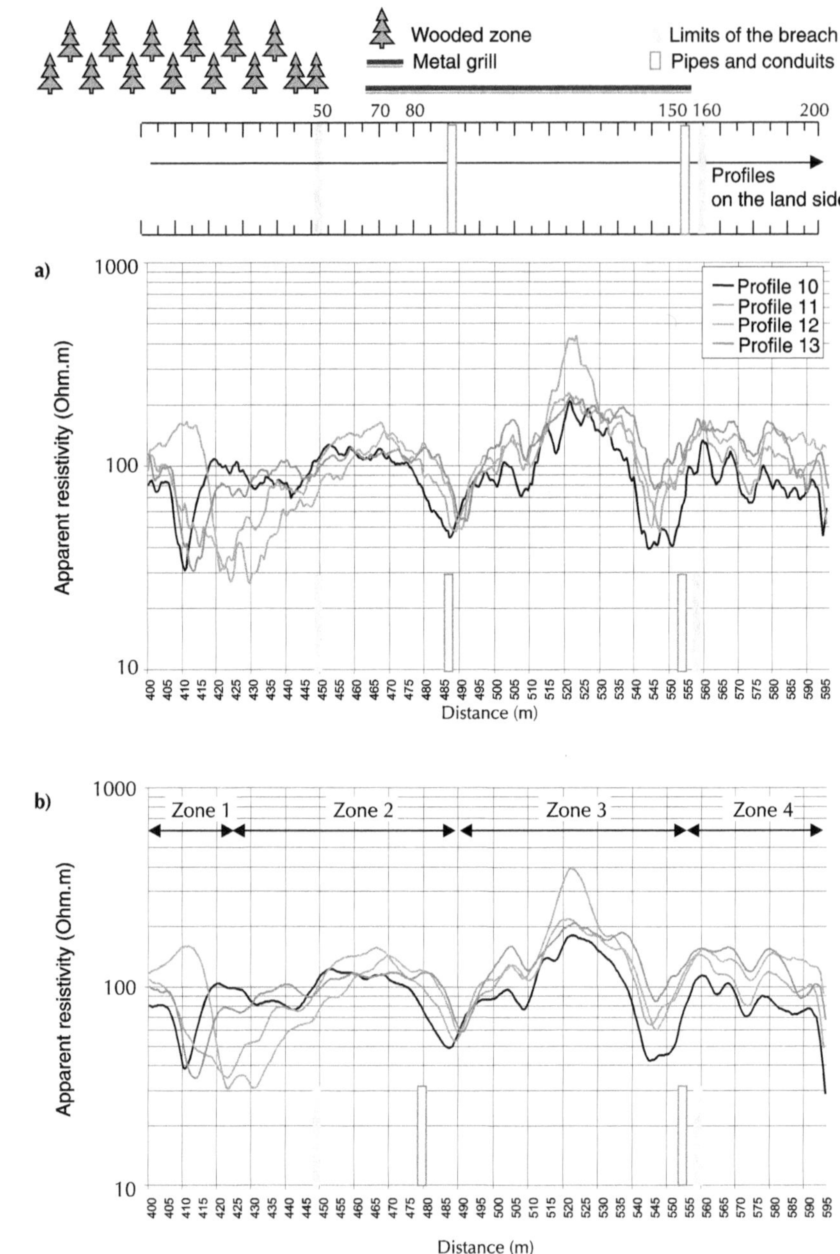

Figure 30 – Comparison of four RMT profiles on the land side. *a)* raw profiles, *b)* filtered profiles

in surface roughness. These problems are sufficient to limit the value of the method when used in dike diagnosis applications at this time. Improvements in sensor technology should be developed.

The results of preliminary studies, and particularly the findings of a visual inspection, are a useful prerequisite to the geophysical measurements. They enable a quantitative interpretation to be made: zones of varying resistivity can be interpreted as zones of varying heterogeneity. Sections where overtopping or breaching occurred are clearly revealed by this method.

VLF methods in inclination mode (VLF-EM or VLF-Z) (McNeill J.-D. and Labson V.F., 1991), whereby the horizontal and vertical components of the total magnetic field are measured, are not described here since they have not been tested. However, since the H field primarily indicates the presence of good conductors, resistivity type methods such as RMT are, at least in principle, the right choice.

4.3 Comparison of Slingram and RMT high-efficiency electromagnetic methods on dry dikes

4.3.1 The efficiency of the methods

The potential of the Slingram method to achieve high levels of efficiency has been demonstrated by investigating the use of a device towed behind a vehicle. The tested device has a reasonable overall size (about 4 m long) and investigates depths of up to about 6 m. Devices, that will necessarily be larger, remain to be developed that could investigate depths of 10 m or more. Alternatively, the devices could have one transmitting coil and several receiving coils set at different intercoil spacings to explore different depths.

In principle, the RMT method offers the best efficiency for dikes that are suited to vehicles. The method's sensitivity to surface heterogeneities and the sensitivity of the capacitive sensors do, however, require that at least three profiles should be determined for any one path, which reduces the efficiency by a factor of three, at least. This shortcoming alone would appear to be sufficient reason to reject this method for dike diagnosis applications. New techniques will have to be developed if RMT is to have a future, particularly to overcome the problems relating to the capacitive mat.

4.3.2 Investigation depth

The investigation depth sounded by the Slingram methods is primarily dependent on the orientation of the induction coils in relation to the ground and on the distance separating them. A large number of devices would provide a simple solution to investigating at the required depths.

The depth that RMT can sound is dependent on the frequency and resistivity of the materials encountered. A sufficient number of transmitters set to between 15 kHz and 1 MHz could easily investigate the body of the dike and some of its foundation.

4.3.3 Measurement repeatability

The RMT method offers good repeatability on roads (Chevassu *et al.*, 1990). However, for dikes, tests have revealed a broad variability in the profiles determined along the same path and only a few minutes apart. At least three profiles must be determined to allow reliable interpretation of the results.

The Slingram method offers very good measurement repeatability, even when comparing a profile produced in walking mode with that produced in towed mode.

4.3.4 Correlations between the two types of measurement

The RMT method detects the apparent resistivity of the materials by measuring the electromagnetic field backscattered by the subsoil. It is sensitive to the presence of both conductive and resistant materials.

The Slingram method measures the apparent conductivity, or its inverse, the apparent resistivity. It is more sensitive to the presence of conductive materials than resistant materials.

Comparative measurements (figure 31) show that the two methods give similar results for an equivalent penetration depth. However, on the Agly dikes (figure 32), comparison of the two methods shows that the breached section is detected and that other resistivity peaks are present on both the Slingram and RMT profiles, but that the RMT measurements reveal other resistant heterogeneities that the Slingram method is less able to detect. The two methods are thus complementary. The two RMT profiles in figure 32, determined at the same frequency along the same path, are also characterised by a large degree of variability for some zones, due mainly to the sensitivity of the method to surface heterogeneities. This again highlights the need to determine several RMT profiles when calculating an average measurement for a given path.

Finally, it is important to note that both the Slingram and RMT methods are effectively measuring the resistivity of a heterogeneous volume of earth and that the result must be considered as an overall, relative value rather than an absolute value.

4.4 Other high-efficiency exploration methods: ground penetrating radar

Ground penetrating radar (GPR), also known as surface penetrating radar or pulse radar, is not the first-choice method for dike investigation work from the crest. It is described here since it can be used at the foot of the dike to study the structure and materials of the foundation. It can also be used on sections of dry dikes so long as the resistivity is greater than 100 Ω.m and the height is less than 5 m – one example being a shoulder of coarse materials or a pavement system.

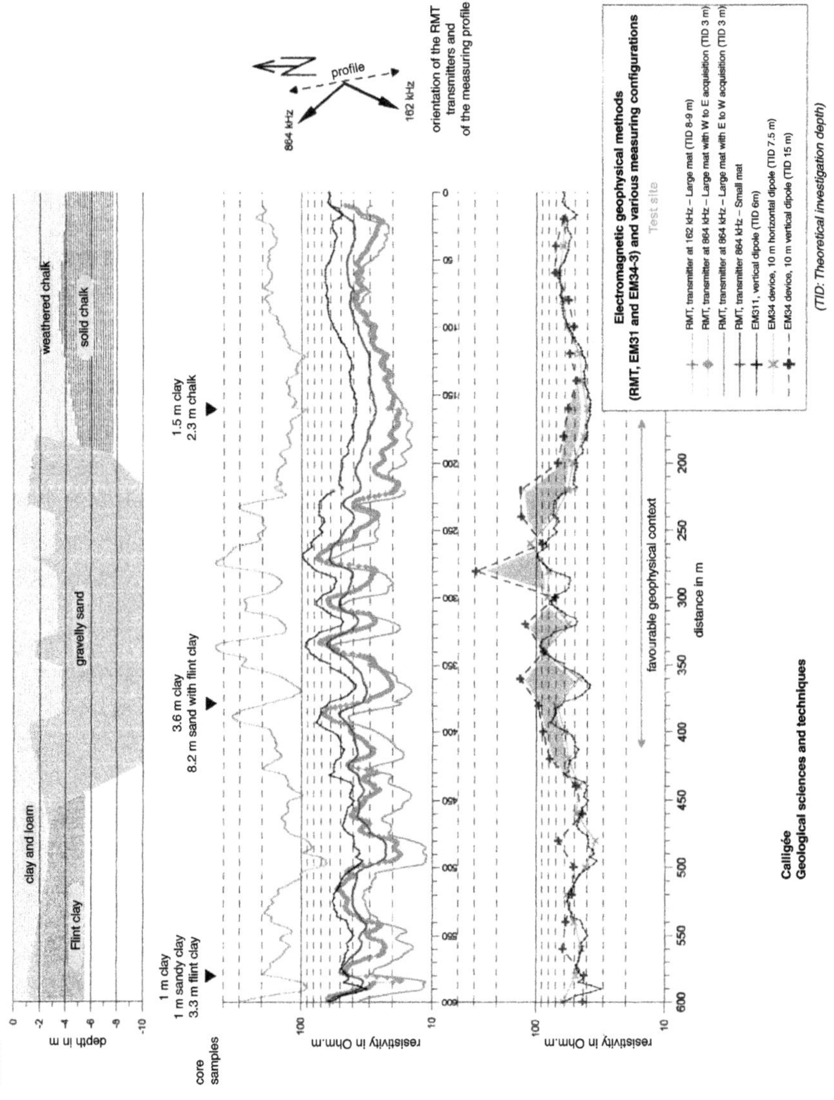

Figure 31 – Comparison of measurements made with the RMT and Slingram methods (Calligée document)

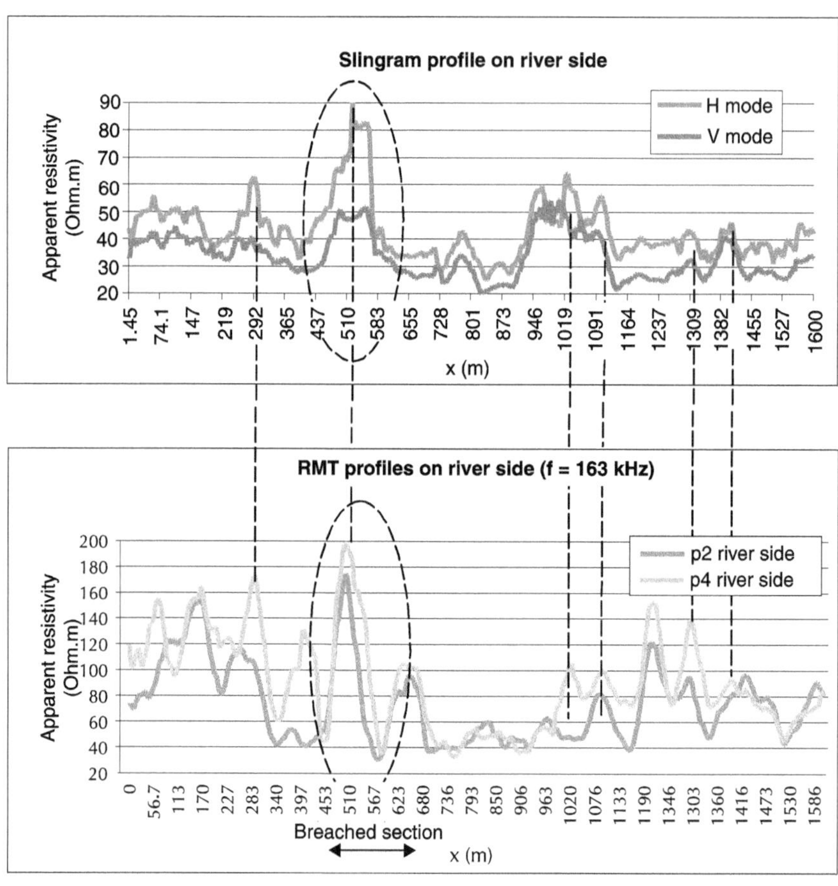

Figure 32 – Comparison of measurements made with the RMT and Slingram methods on the Agly dike

4.4.1 Principle

Explorative radar methods (Daniels *et al.*, 1988; Daniels, 1996; Davis and Annan, 1989) are based on analysing the propagation of electromagnetic waves in the soil within a frequency band ranging from a few dozen MHz to a few GHz. These waves are transmitted in the form of very short pulses from a point on the surface by a transmitting antenna. They interact with the subsurface material when they encounter a contrast in the electromagnetic permittivity and are partially reflected back to the surface where their characteristics are measured by an antenna and analysed so as to deduce the properties of the subsoil (figure 33). In monostatic mode, the antenna performs the dual role of receiver and transmitter. In bistatic mode, there is one transmitting antenna and one receiving antenna. Devices with a much larger number of antennas could potentially be developed (multistatic mode).

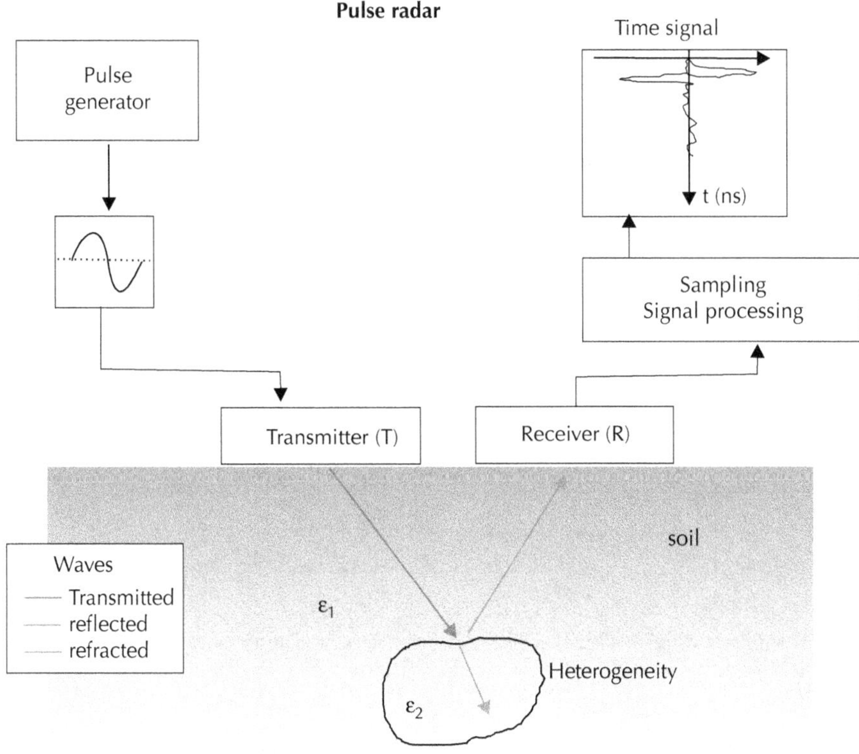

Figure 33 – Principle of the ground penetrating radar

The quantity measured is a signal that indicates amplitude variations in the electric field measured as a function of the propagation time of the waves in the ground (measured in nanoseconds).

4.4.2 Output

The first expected result is called the **raw radargram** (or time cross-section): the amplitude of each signal or scan is graduated into a scale of colours and the signals are juxtaposed according to their position on the surface.

The second expected result is the **interpreted radargram:** if the speeds of the signal in the media are known, then these signals can be shown in terms of depth (m). Juxtaposing several radargrams (figure 34) produces a 3D image of the measurements (explored surface in the horizontal axis and the depth in the vertical axis) using a cross section at a given depth or along a given profile.

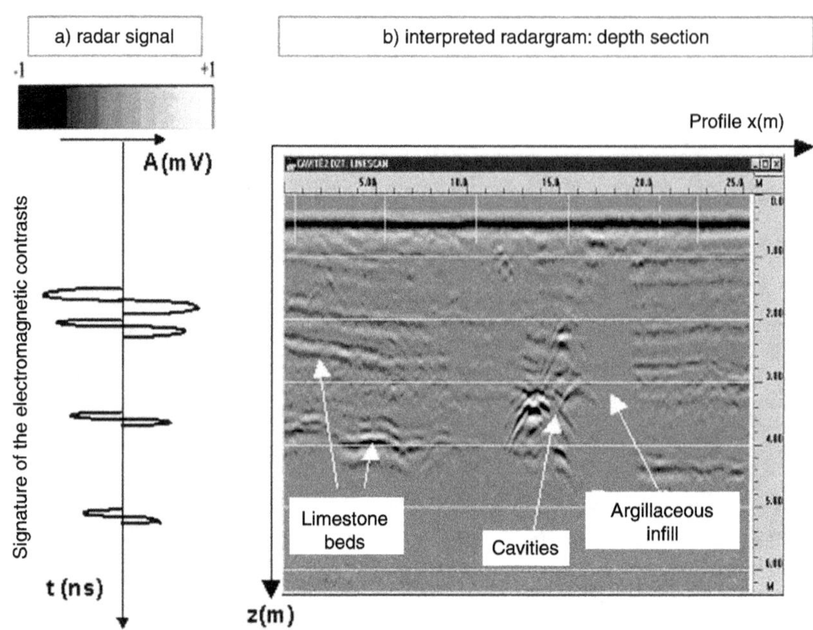

Figure 34 – *a)* Radar signal or scan, *b)* example of an interpreted radargram produced at a centre frequency Fc = 200 MHz (LRPC document, Saint-Brieuc)

Interpreting the raw radargrams for a dike is therefore dependent on determining the speeds of the waves in the materials through which they pass. Assuming a far-field context, at a given frequency, an electromagnetic wave produced by the transmitting antenna is a harmonic plane wave. Figure 35 illustrates the electric and magnetic fields in space for this plane wave.

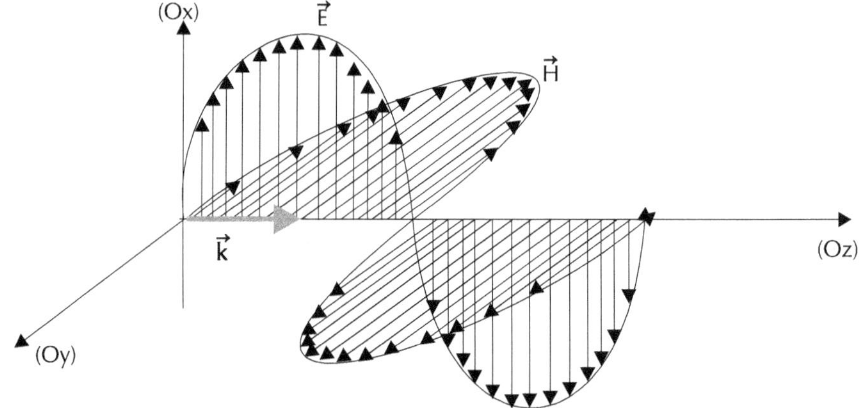

Figure 35 – Illustration for a harmonic wave of the electric and magnetic fields in the direction of the $\vec{k}$ vector, the propagation vector, which in this case is the (Oz) direction

The wave is planar if all the points in a plane perpendicular to the direction of propagation experience fields that are equal in amplitude and phase. The electric field $\vec{E}$ and the magnetic field $\vec{H}$ are, in the plane wave, perpendicular to the propagation direction (Oz).

For a monochromatic wave with an angular frequency ω propagating in an axis (Oz), that is polarised in an axis (Ox) of unit vector $\vec{j}$, in a medium whose effective permittivity is $\varepsilon_e = \varepsilon_e' - j\varepsilon_e''$ (where $\varepsilon_e' = \varepsilon' + \sigma''/\omega$ and $\varepsilon_e'' = \varepsilon'' + \sigma'/\omega$, where ε' and ε'' are the real and imaginary parts of the permittivity and where σ' and σ'' are the real and imaginary parts of the conductivity), the general solution to the Helmotz equation can be written as:

$$\vec{E}(x, y, z, t) = E_0 e^{-jkz} e^{jwt} \vec{j}$$

where k is the wave number defined as:

$$k = \beta - j\alpha$$

where α (in np/m) is the attenuation coefficient of the wave and β (in rad/m) is the phase coefficient.

The velocities in the medium can be calculated if the permittivity and conductivity are known. Since most of the soils encountered are not magnetic, the magnetic permeability is taken to be the permeability of free space.

In many cases dealt with in geophysics, the media are considered to be non-dispersive (ε_e is independent of the frequency), within which propagation phenomena dominate and the losses are small ($\varepsilon_e = \varepsilon' = \varepsilon_r' \varepsilon_0$), **which is the case on the surface of dikes on which a road has been built.**

If conduction phenomena are present, **the losses in the medium will be greater. This is often the case in the body of a dike**. The characteristics for a low-loss and a high-loss medium are summarised in Table 3 on the following page.

4.4.3 Modelling

Modelling the propagation of radar waves can help to interpret the measurements. There are several methods, the most common of which are the method of moments and the FDTD method (*Finite Difference Time Domain*). Moreover, a number of software programs have been commercialised by geophysicists and universities that rapidly create a model of the explored medium.

4.4.4 Factors to consider – Methodology for the exploration

Radar-based exploration of dikes is generally limited to the shallow subsurface, although it does clearly reveal contrasts between gravelly and resistant materials. The method offers very good rates of ground coverage (several kilometres to several dozen kilometres per day). Depending on the equipment used and the conditions, the device may be carried by the operator or dragged behind a vehicle.

Medium	Low losses (road built on dike, draining shoulder)	High losses (loamy or clayey dike body)
speed v (m.s⁻¹)	$v = \dfrac{c}{\sqrt{\varepsilon'_r}}$	$v = \dfrac{\omega}{\beta} = \dfrac{c}{\sqrt{\dfrac{\varepsilon'_r}{2}} \cdot \sqrt{1 + \sqrt{1 + \tan^2\delta}}}$
attenuation cœfficient α (Np.m⁻¹) and phase cœfficient β (rad. m⁻¹)	$\begin{cases} \alpha = \dfrac{60\pi\sigma'}{\sqrt{\varepsilon'_r}} \\[2ex] \beta = \dfrac{\omega}{c}\sqrt{\varepsilon'_r} \end{cases}$	$\begin{cases} \alpha = \omega\sqrt{\dfrac{\varepsilon_0\varepsilon'_r\mu_0}{2}\left(\sqrt{1 + \tan^2\delta} - 1\right)} \\[3ex] \beta = \omega\sqrt{\dfrac{\varepsilon_0\varepsilon'_r\mu_0}{2}\left(\sqrt{1 + \tan^2\delta} + 1\right)} \end{cases}$
loss angle $\tan\delta$	$\tan\delta = \dfrac{\sigma'}{\omega\varepsilon'_r}$	$\tan\delta = \dfrac{\varepsilon''_r + \dfrac{\sigma'}{\omega\varepsilon_0}}{\varepsilon'_r} = \dfrac{\varepsilon''_e}{\varepsilon'_e}$
penetration depth d (m)	$d = \dfrac{\sqrt{\varepsilon'_r}}{60\pi\sigma'} = \dfrac{\rho\sqrt{\varepsilon'_r}}{60\pi}$	$d = \dfrac{1}{\alpha} = \dfrac{1}{\omega\sqrt{\dfrac{\varepsilon_0\varepsilon'_r\mu_0}{2}\left(\sqrt{\sqrt{1 + \tan^2\delta} - 1}\right)}}$

Table 3 – Equations for some characteristic quantities for high-loss and low-loss media

However, at depths greater than 2 to 3 m, the quality of radar data drops significantly: the component materials of the dike body tend to be conductive (< 100 Ω.m), and instead of penetrating the material, the radar waves are absorbed by it.

The measuring interval between profiles depends on how much detail is required. It varies from 0.5 to 5 m or more, depending on the purpose of the investigation and the size of the zone under consideration.

The **choice of centre frequency** for the survey is important. It is determined to a certain extent by the characteristics of the antenna as well as by two criteria. The first criterion is the operating bandwidth: the wavelengths used must be about the same size as the target, and the bandwidth must be wide so as to provide good temporal resolution. The second criterion is the nature of the ground and the depth of interest, which is rarely more than twenty metres in resistive ground (*i.e.* a resistivity of more than about 100 Ω.m).

The **signal recording time** is set based on the information collected during the search for indicators. If the speed of the waves in the soil, and the depth of the targets being investigated are known roughly, then it is possible to estimate the propagation time of the waves to the target. The length of the recording time will be double this time, corresponding to the two-way travel time of the waves to the target (and known as the time window).

The operator must then set a **signal recording increment**. This is often adjusted using a dial that triggers an acquisition point each time a given distance increment is reached. Data may also be acquired continuously, in which case the operator must walk at constant speed and indicate points regularly along the profile, whose coordinates are known.

Other parameters may be set:

– **stacking:** this option sums several signals measured at exactly the same point, so as to improve the signal-to-noise ratio;

– **gain:** commercially-available radars generally offer the option of applying gains to amplify the signal as a function of the listening time;

– **filters:** it is generally possible to apply bandwidth filters to the data being acquired. If this technique is used, their characteristics should be known so that this can be considered during the post-acquisition processing.

Finally, and as with all the methods involving measurements made directly on the surface, a **survey of the topography** is valuable for resetting the signals to the datums. Consider nearby structures, such as high-voltage power lines, buildings with metal walls, buried cables, radio transmitters, etc. which, depending on the quality of the shielding of the antennae, can generate reflections in the signals recorded and may result in interpretation errors.

4.4.5 Some features of the measuring equipment

Ground penetrating radar devices transmit, receive and record electromagnetic signals. The signals sent are pulses with a time duration ranging from about one nanosecond (for studying roadways or searching for steel elements in concrete) to a few tens of nanoseconds (geological applications). The pulse repetition frequency may be as high as several hundred kHz. The reflected signals (or *scans*) are recorded during a period of time known as a time window or range. The latest devices can record several hundred signals per second, with a time resolution of about 5 picoseconds. The dynamic range of the radar devices is about 110 dB. This equipment processes the received signals in real time (bandwidth filters and signal summing). Physically, the radar devices are box-like units with an associated screen, which may or may not be attached to the unit, used to read the measured signals in real time.

The transmitting and receiving **antennas** used for GPR devices vary greatly, although several different types may be used for the same application. The antenna characteristics, how they operate and their electromagnetic radiation are important considerations that can be compared for different devices, for example in Combes, 1996 and Eyraud *et al.*, 1973. The antennae are generally cylindrical dipoles for low-frequency applications, bowtie or butterfly dipoles for intermediate applications and some high frequency applications, and horns for use on roads.

With the exception of horn antennas, GPR antennas operate in contact with the ground, or a few centimetres above it, which can cause problems in terms of knowing their radiation pattern in the ground under investigation. They are usually shielded for protection from above-ground electromagnetic noise.

Characteristic antenna features include:

– a **centre frequency of radiation**, *i.e.* the frequency at which the transmitted energy is maximal;

– a **bandwidth**: the usable frequency range around the centre frequency;

– a **radiation pattern**: a graphical representation of the spatial distribution of radiated energy in terms of angle from the principal radiation direction;

– a **half-power beamwidth (– 3dB)**: the angle in one of the polarisation planes (plane E or H) with the principal radiation direction in which the transmitted power is one-half of the peak power of the antenna pattern.

4.4.6 Interpreting the measurements

The **raw measurements** provide **qualitative information** for locating anomalies directly in the ground by the identification of a contrast or attenuation in the recorded signal. When the speeds are not known, geotechnical surveys can be conducted to characterise the soil structure in terms of permittivity and depth.

For optimum resolution, the frequency of the electromagnetic waves should be high. However, at high frequencies the penetration depth of the waves is very poor. If the frequency is reduced to improve the penetration, the resolution suffers.

If the speeds in the ground are known, a **quantitative interpretation** of the radargrams may be carried out in terms of depth. The dimensions of the anomalies may also be evaluated by determining the **vertical resolution**, estimated to be about one quarter of the wavelength corresponding to the centre frequency in the material, $r_v = \lambda/4 = \dfrac{1}{4}\dfrac{c}{\sqrt{\varepsilon'_r}}\dfrac{1}{f_c}$ (m) and the **horizontal resolution** defined for a depth z by: $r_h \approx \sqrt{\dfrac{\lambda^2}{16} + \dfrac{\lambda z}{2}}$ (m). Irrespective of the result of a quantitative interpretation, it is advisable to conduct a thorough geotechnical survey of the anomaly.

4.4.7 Example of results

The breached section on the Agly dike, described in the section on Slingram and RMT methods, was investigated using GPR (Mériaux *et al.*, 2003). Three radar runs were performed by two operators in one day in the field in the Saint-Laurent breach sector (on either side of the crest roadway and in the centre), as well as fifteen transverse profiles. The radar used was an SIR-10A+ manufactured by GSSI, with an antenna operating at 500 MHz. The repaired breach was identified clearly by the presence of numerous diffraction points (figure 36). Roadway structure materials (about one metre deep) were found to be more absorbent downstream of the breach than upstream.

As indicated in figures 28 and 18, the Saint-Laurent breach appears to be a resistant anomaly. Indeed, it was filled with coarse and resistant materials. This zone is thus, in principle, more permeable to radar waves than the rest of the body of the dike; a finding that is confirmed in figure 36.

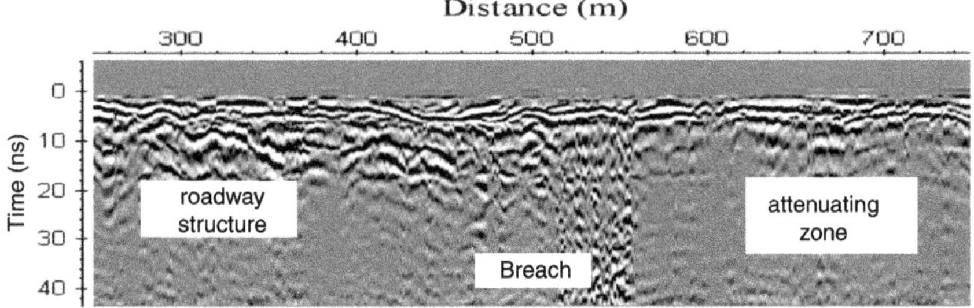

Figure 36 – Radargram revealing the breach, GPR at 500 MHz along the roadway (Mériaux *et al.*, 2003)

4.4.8 Conclusion

GPR is a high-efficiency method that **is poorly suited to most dike diagnosis**. Radar waves do not penetrate the conductive materials (loams, clays, etc.) that are the predominant dike body components. Consequently, the penetration depth is generally limited to the first two metres below the crest of the dike, particularly if there is a road along the crest. The method should not, however, be rejected outright for this application: certain dike sections are compatible with the use of radar (*e.g.* a dike with a resistant structure (> 100 Ω.m) and less than 5 m high), or resistant parts of dikes, such as a breach repaired with coarser materials (roadway structure, draining shoulder, etc.). Moreover, where the reservations mentioned above are overcome, it offers a simple means of producing transverse profiles.

4.5 Local investigation with 2D electrical imaging survey

This method provides a valuable complement to high-efficiency measurements, or can be used as a follow-up to geotechnical tests. The type of geoelectrical method described below, the 2D electrical imaging survey, is particularly effective at creating a detailed image of a transverse or longitudinal section of the dike. It maps the distribution of heterogeneities (by measuring their resistivity) as a function of (approximate) depth, whereas high-efficiency methods produce a profile that reflects the overall distribution of the materials in a section of the body of the dike, with no real precision regarding the depth.

4.5.1 Principle of electrical imaging survey methods

Direct current electrical imaging survey methods determine the properties of soils by measuring their resistivity ρ (Ω.m). The measuring principle is as follows: a direct current I is injected into the ground by two electrodes, A and B. The potential difference V is measured at the terminals of two other electrodes, M and N (figure 37).

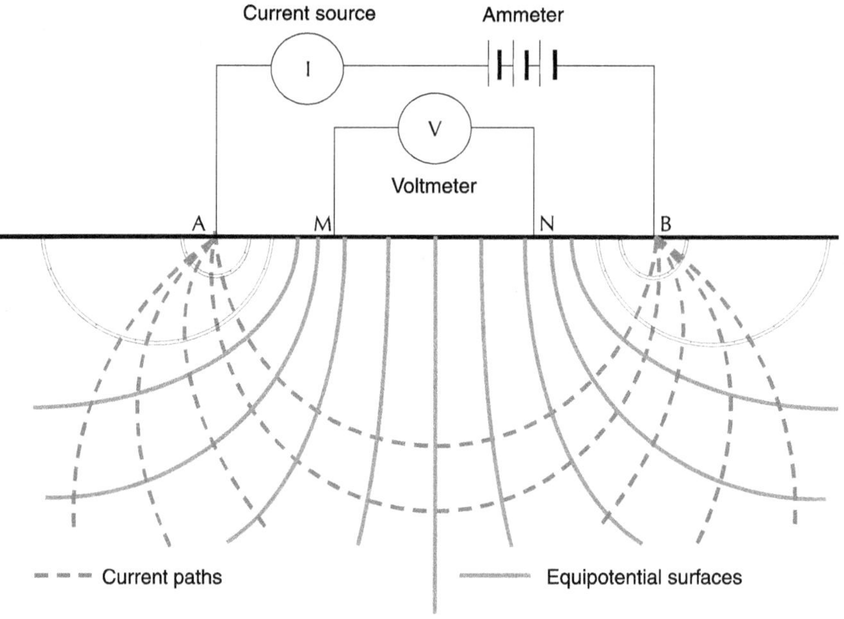

Figure 37 – Principle of direct current electrical imaging surveying

The apparent resistivity is "the ratio of the potential difference measured in the ground to that which would be measured by the same device for the same injection of current into a homogeneous medium with a resistivity of 1Ω.m". This quantity has the dimensions of resistivity. It has been shown to be equal to:

$$\rho_a = k\frac{V_M - V_N}{I}$$

where k, the geometric factor, is defined by:

$$k = 2\pi\left[\frac{1}{AM} - \frac{1}{MB} - \frac{1}{AN} + \frac{1}{NB}\right]^{-1}$$

The most popular devices are shown in figure 38. These are all quadripoles. The most commonly used devices are the Wenner (alpha and beta), Schlumberger, dipole-dipole, pole-pole and, increasingly, the pole-dipole in an array. Their investigation depth increases as the distance AM is increased. For all the devices, this distance is defined as being the smallest distance between the current injection electrode and the potential electrode. If the distance AM is set up to be the same for all the devices, then all the devices will offer an equivalent depth of investigation. Authors such as Militzer *et al.* (1979) and Roy & Apparao (1971) have studied in more detail the depth of investigation offered by the various devices. As a first approximation, the depth of investigation may be considered to be approximately $1/6^{th}$ to $1/8^{th}$ the length of the device.

There are three ways of carrying out direct current electrical surveying from the surface:

– **vertical electrical sounding**: the centre point is fixed on the surface, and the spacing between the electrodes is increased;

– **electrical profiling**: the device is moved along the survey line, with the spacing between the electrodes fixed;

– **2D electrical imaging survey**: a combination of the above two methods.

The methodology is described for the 2D electrical imaging survey and is equally applicable to the two other methods.

a) Vertical electrical sounding

For the vertical electrical sounding device, the centre point is fixed and the electrode spacing is increased so as to measure the apparent resistivity as a function of the length of the device. This principle is illustrated in figure 39 for a Wenner device, with the distance between the electrodes at points A, B, M and N varied by an integer number of times the initial separation a.

The **quantity measured** is the apparent resistivity of the ground below the centre of the device: vertical electrical sounding provides information about vertical variations in the resistivity of the ground. The depth of investigation is determined by the final measuring device length and is limited by the resistivity of the subsoil.

The **output** is a logarithmic curve showing the apparent resistivity (Ω.m) as a function of the distance between current injection electrodes (generally expressed in metres, figure 39).

For stratified media, the results are interpreted in terms of the thickness and resistivity of the layers. Charts are used to determine the apparent resistivity of multilayer media (Parasnis, 1986). More recently, this process has been simplified by computerised interpretation software. The interpretation does not always yield unique solutions, since the laws of similarity are applicable to this processing.

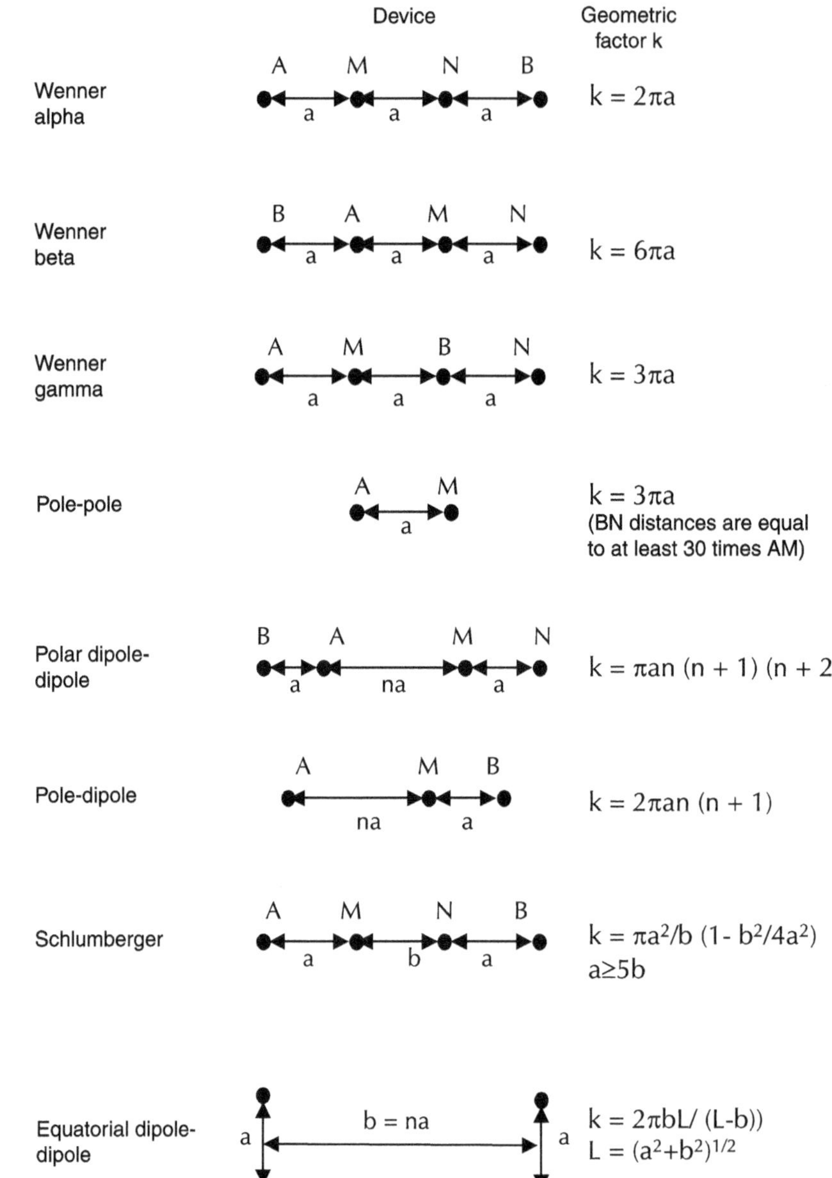

Figure 38 – Common arrays and their geometric factors (Loke, 2002)

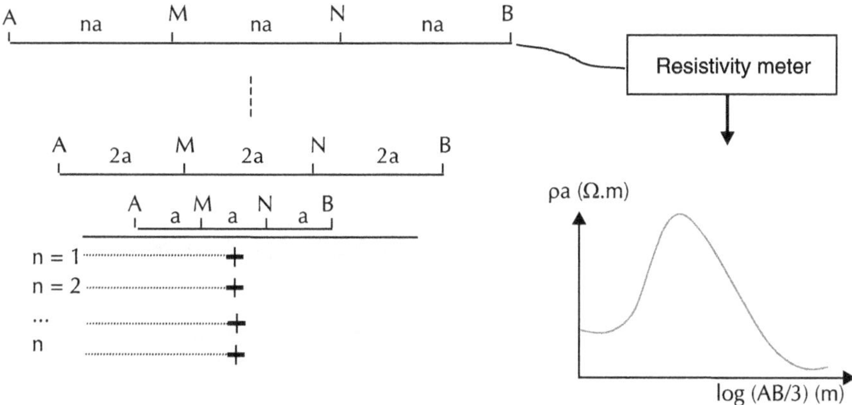

Figure 39 – Principle of vertical electrical sounding using a Wenner device

b) ELECTRICAL PROFILING

With this method, a device with fixed electrode spacing is dragged along the profile, with the electrodes in line with the profile. The principle is illustrated in figure 40.

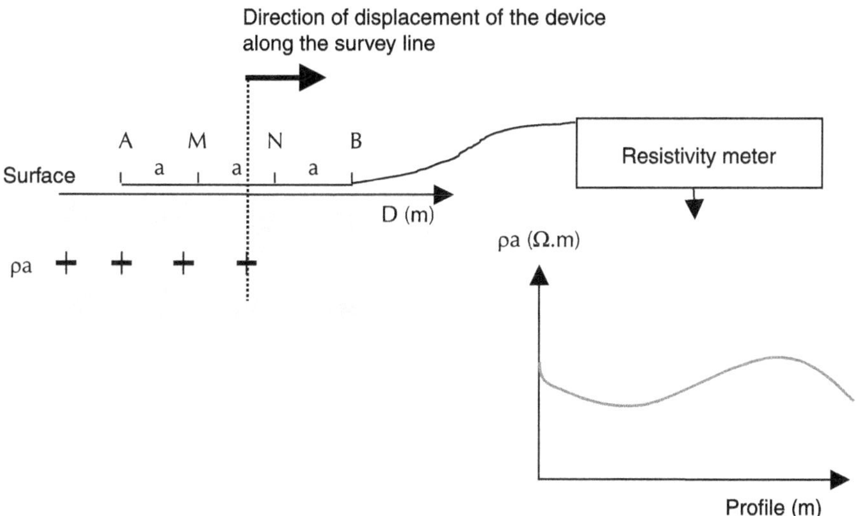

Figure 40 – Principle of the electrical profiling, for a Wenner device

The **measured quantity** is the apparent resistivity of the soil for a given device length. The depth of investigation is dependent on the length of the device, and on the resistivity of the subsoil. This device is particularly sensitive to lateral variations in resistivity. Several runs along a profile can be collated to produce a map of the apparent resistivity.

The **output** of an electrical profiling survey is a semi-logarithmic curve showing the apparent resistivity (in Ω.m) of the ground as a function of the position of the measuring device along the profile (in metres) (figure 40).

4.5.2 2D Electrical imaging survey

a) PRINCIPLE

This method combines the principles of the electrical profiling with vertical electrical sounding. We strongly recommend reading the tutorial given by Loke (2002), in which data acquisition, presentation and arrays are explained in depth.

A large number of electrodes (for instance 64, 96 or more) are inserted at constant intervals along a rectilinear profile (2D) or in a grid pattern (3D). All the electrodes are connected to a potential measuring and current injection device, alternating between injection electrode and potential measurer. The system (current injection and potential measurement) is computer-controlled and the sequence of measurements to take, the type of array (Wenner, pole-pole, pole-dipole...) and other survey parameters are defined.

For instance, the Wenner electrode array is briefly described here in figure 41: the first measurement is made with the two current electrodes A and B, and the two potential electrodes M and N, with a separation of "1a" from each other's. The second measurement is made by advancing the four electrodes by "1a". After completing a sequence spacing "1a" is achieved, the measurement with "2a" spacing is performed. Measurement is complete when the "na" spacing has been performed.

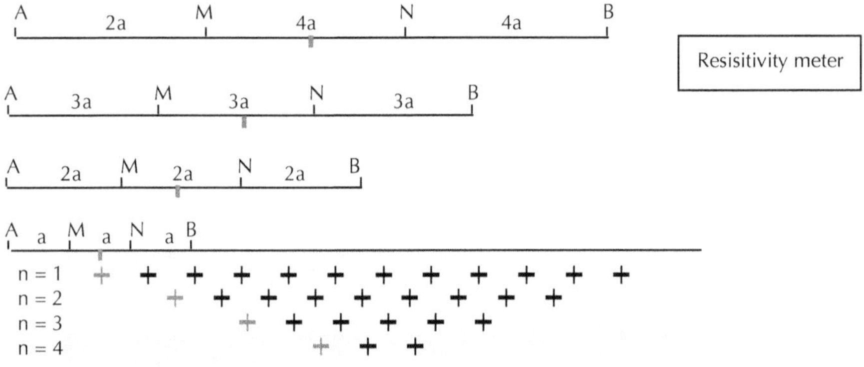

Figure 41 – Principle of 2D electrical imaging survey

b) OUTPUT

The raw output of 2D electrical imaging survey measurements is a map of apparent resistivity (Ω.m), often called a **pseudo-section.** The position of the centre of the device along the profile is plotted on the x axis (in metres) and the device length is plotted on the y axis (in metres). The result is not a real representation of a section through the ground. For a given y-axis value (*i.e.* device length), the map reveals the apparent resistivity for this device length along the profile. However, the apparent resistivity is affected by any heterogeneity located in a subsurface hemispherical volume lying below the centre of the device. Interpreting the raw output is very difficult and requires a great deal of experience.

Processing the measurements produces an **inversion resistivity map (Ω.m)**. Software is available that inverts the measurements so as to illustrate variations in the interpreted resistivity as a function of the position of the device and the depth. This software assumes a two-dimensional distribution of the resistivity of the component materials of the subsoil, and despite the fact that the method considers the entire hemispherical volume located under the device. Consequently, a heterogeneity revealed by the inversed results may very well be caused by an anomaly on the surface, and not one at depth. The results of the inversion should therefore be considered with care.

Modelling reveals that small heterogeneities on the surface can mask deep and large heterogeneities. The system also picks up noise from geological sources, occasionally at high amplitudes. This method does, however, provide a wealth of information that is useful in refining the model of the internal structure of the dike (stratification, dips and faults).

c) Methodology

A check (by vertical electrical sounding) should be carried out to confirm that the measured apparent resistivity varies with the length of the electrical device. The results of this test are used to make any necessary adjustments to the device lengths, which should be about six to eight times the target depth. Modelling may help in setting the parameters of the device. For depths of less than 20 m, the pole-pole device should be preferred. Only the A electrodes (current) and M electrodes (potential) can be moved, electrodes B and N are at a distance of more than twenty times AM and must be located at an angle of 30° to the AM axis. The symmetrical quadripole device is used for greater depths, as is the pole-dipole device. The dipole-dipole device is often used for 2D imaging survey, but can produce results that are difficult to interpret, since it is highly sensitive to background measurement noise.

To overcome the interpretation difficulties associated with the 3D location of heterogeneities, several parallel profiles can be determined (transverse or longitudinal). Access to a topographic survey is required for the processing of the transverse measurements.

The device settings must be kept as constant as possible, as recommended in the AGAP's Code of Good Practice (see the bibliography). Particularly for a rectilinear device, the direction should not deviate by more than 10 degrees from the direction of the reference profile. The electrode spacing must be known to the nearest 5%. Finally, the measurement of the V/I ratio must be known to the nearest 3% and the measurement of the potential difference between the V electrodes must be greater than 0.5 mV, since the potential difference is particularly sensitive to electrode polarisation phenomena and to currents in the ground (telluric currents).

d) INTERPRETING THE RESULTS

The raw results are generally difficult to interpret: the geometry of the indicated anomalies very often differs from that of the actual heterogeneities. The results can only be interpreted by considering inversion profiles. Great caution must be exercised when suggesting the accurate location of deep anomalies, since the image produced is the sum of all the heterogeneities located in the subsurface hemisphere under the point of investigation. Moreover, there is not a unique solution for a measured anomaly. Finally, the actual investigation depth varies; a factor that the interpreted investigation depth does not take into consideration. However, the section does provide valuable information about the contacts between layers and the geometry of the dike, which geotechnical soundings can then refine into absolute values.

e) EXAMPLES OF RESULTS

The plot below (figure 42) shows measurements made with a 2D electrical imaging device along a longitudinal profile of a dike on the Isère river (France). The raw interpretation indicates that the dike is made of materials of varying resistivity. Following a geotechnical sounding, the geological interpretation can begin: red and orange zones correspond to gravelly fill with higher resistivity, which is distributed regularly along the dike. The blue shades correspond to the clay loam base of the dike. The yellow and green zones correspond to sandy materials. At 36 m along the profile, there is clearly a resistant anomaly. The vertical scale gives an indication only of the interpreted depth of the anomalies. Only soundings can determine the true depths.

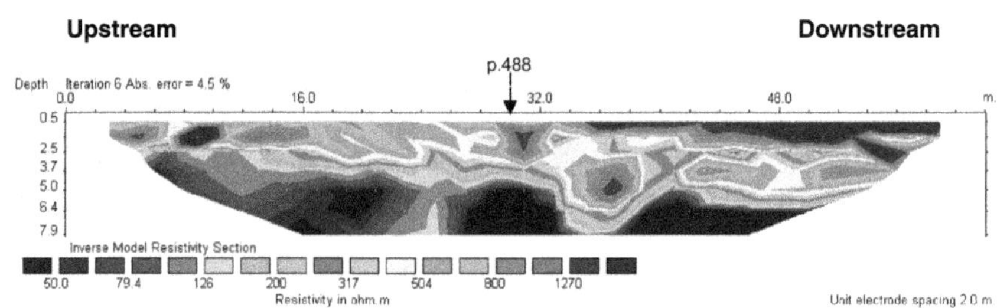

Figure 42 – Example of 2D electrical imaging measurements along a section of the dike on the Isère river (IMS document)

Another example of 2D electrical imaging survey measurements is given in figure 43. This transverse profile illustrates a classical geometry of a dike body that has been integrated into a river bank. The fill material, gravelly and more resistant (shown in orange and red for the coarsest material, then yellow and green for the finest), is thin on the land side (fields) and thickens on the river side up to the berm. The bed of the dike is made of more compact clay loam materials and is shown in blue. The resistant anomaly on the land side, detected on several transverse and parallel profiles, is actually a buried conduit.

Figure 44 compares the results from a 2D electrical imaging survey with those from an EM31 (Slingram-type) device for a dike on the Cher river (shown in section in figure 21). Note that the resistivity scale is not logarithmic. The zone of low resistivity revealed by the 2D electrical imaging is identified similarly in the EM31 profile at around metre marker (MM) 6290, and the same is true for the more resistant zone that precedes it. The 2D electrical imaging also reveals a near-surface heterogeneous zone between MM 6295 and MM 6340, consisting of a conductive top slab about 5 metres thick, corresponding to repair work carried out after the breach of 1856 and backfill that was more loamy than the surrounding materials. This zone lies above a more resistant layer, starting about ten metres down, corresponding to alluvium of coarse sands and to a chalky calcareous substratum (figure 21).

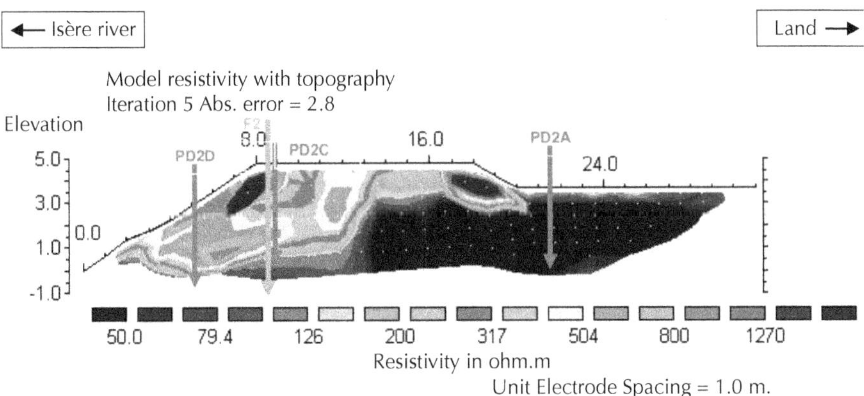

Figure 43 – Example of 2D electrical imaging measurements of a transverse profile of a dike on the Isère river. (IMS document)

This example clearly illustrates the relative strengths and weaknesses of low-frequency electromagnetic methods and array-type direct current electrical methods. The first is a high-efficiency method. It gives an apparent value of the resistivity of an investigated volume as a function of position on the surface. The method is straightforward and the results are relatively easy to interpret. The second method offers a low rate of ground coverage (a few hundred metres per day). Whilst it is similar in that it gives an apparent value of the resistivity of an investigation volume as a function of position on the surface, this information is also provided as a function of the interpreted depth of the heterogeneity.

f) Conclusion

2D electrical imaging is an effective method since it can explore the entire body of the dike. Its rate of ground coverage is low (a few dozen metres per day) since the device takes a long time to set up. However, the measurements produced in the form of resistivity maps are useful in showing the distribution, at a suitable resolution, of the materials in the body of the dike. Moreover, the maps can reveal different zones of materials, *e.g.* zones of high resistance corresponding to coarse materials (more gravelly materials) used to backfill or weight the dike can be differentiated from the original materials of the dike body, that are generally finer (less resistant materials corresponding to sandy and clay loam materials). The method can also produce transverse profiles that reveal the overall geometry of the dike.

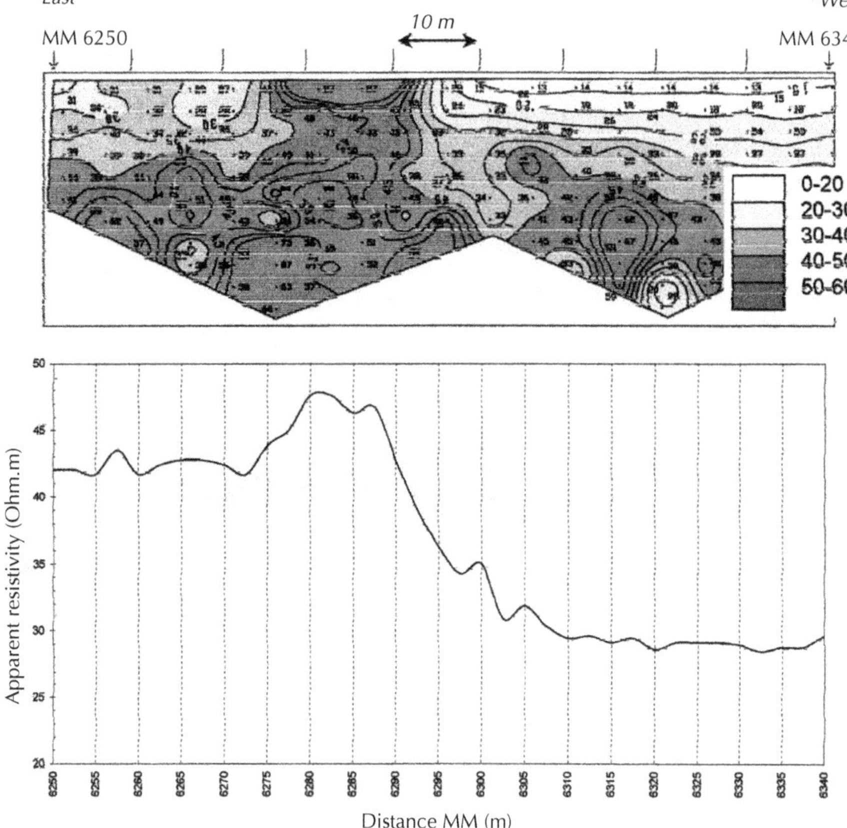

Figure 44 – Comparison of measurements made with a 2D electrical imaging device and an EM31 (Slingram-type) device (Frappin, 2001)

g) Footnote. High-efficiency but less effective method: 2D capacitive electrical imaging

The principle is the same as that described above. However, this electrical method does not require the insertion of electrodes into the ground. These devices drag an array of capacitive sensors over the ground, offering much greater ground coverage since they can be towed by a vehicle. For the device tested by the authors, the depth of investigation was low – about 2 to 4 m, depending on the resistivity of the subsoil and on the electrode spacing. The nature of the device limits its use to flat surfaces covered with short grass.

An example of the measurements produced is shown in figure 45. In view of the shallow investigation depth, this device is primarily used to investigate the fill material along the crest of the dike. As with the inserted 2D electrode system, the results must be interpreted with care since they provide a two-dimensional representation of a three-dimensional distribution of heterogeneities. The information that can be gained is therefore limited. However, it is clear from profiles B and C that, for the first 20 metres, the fill appears to consist at the surface of a succession of gravelly (resistant) and loamy materials (conductive) and that after the 20 m point, there is a greater proportion of more resistant materials in the estimated 2.5 m of investigation depth.

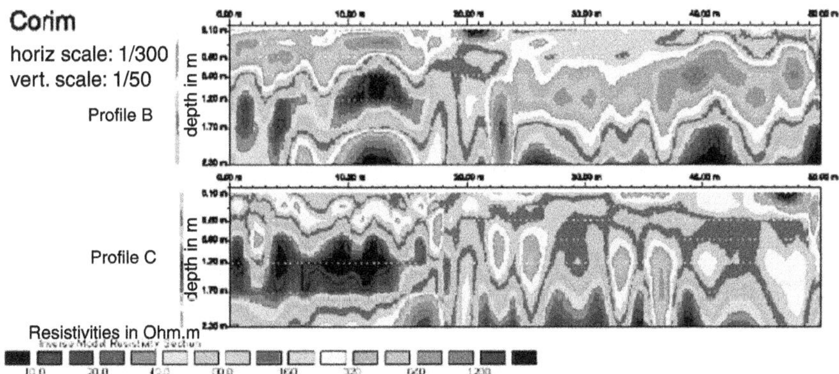

Figure 45 – Example of measurements made with a 2D capacitive electrical imaging: two profiles along the crest of the dike on the Isère river (IMS document)

The 2D capacitive electrical imaging array allows a greater rate of ground coverage for dikes whose crest can be driven along, although the information collected, for the device tested, is limited to investigating the near-surface materials of the dike. Moreover, the correlations with the 2D electrical imaging method suggested by the results were poor. Greater investigation depths, of about a dozen metres, would require device lengths of about 60 to 80 m. The physical difficulties of using such long devices along non-rectilinear profiles would appear to rule out this method; this is not the case for low-frequency electromagnetic methods.

4.6 Local investigation using seismic refraction

4.6.1 Principle

Seismic investigation methods conducted at surface level are based on analysing the propagation of mechanical waves in the soil. These waves are generated at a point source by a single or repeated impact (or vibration). If the waves encounter a contrast in mechanical impedance, they are partially reflected and this wave is picked up by a receiver (known as a geophone). Measurement and analysis of the characteristics of the refracted wave provide information about the properties of the subsoil (figure 46).

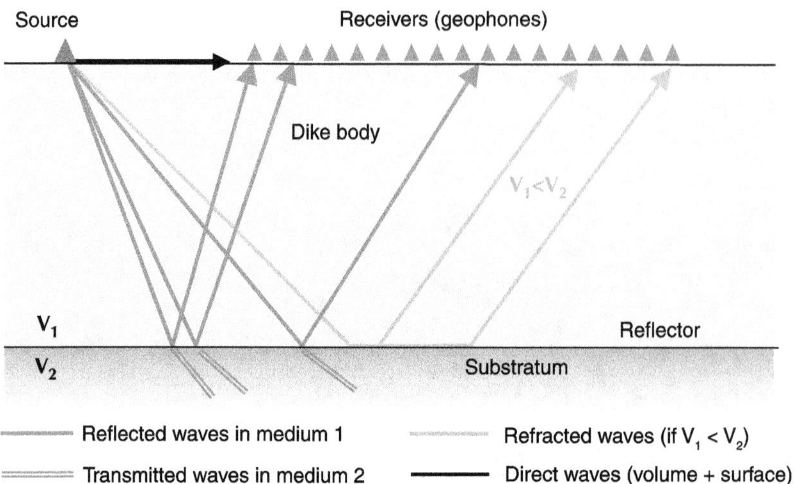

Figure 46 – Principle of seismic methods

The laws of propagation of seismic waves are similar to the laws of optics Snell and Descartes' laws of reflection and refraction, figure 47). Their paths may be complex. The travel time is dependent on the path taken.

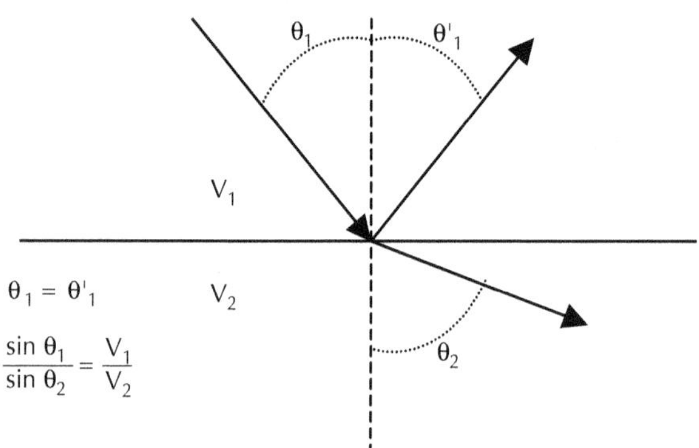

Figure 47 – Snell and Descartes' laws

Three seismic methods can be applied to dike diagnosis:

– high-resolution seismic reflection;

– surface-wave seismic systems – a method currently being developed;

– **seismic refraction**, which has already been applied to dike diagnosis. This is the method described below.

Seismic refraction is the study of the propagation of seismic waves refracted in the soil. As illustrated in figure 48, in a multi-layer medium in which the velocity through each medium increases as a function of depth ($V_2 > V_1$), there is a critical angle θ_c above which the waves are refracted: the path of the wave then follows the interface between the layers and the waves are partially transmitted to the surface at the same critical angle.

This method is applicable to stratified media, where the velocity through the media increases as a function of depth; this is often the case since the degree of compaction generally increases with depth. In a dike diagnosis context, it is commonly used to investigate and monitor changes in the position of the bedrock under the foundation of a dike.

4.6.2 Quantities measured

The quantities measured are presented in the form of **raw seismograms**. For a source lying on the profile, they show, in the vertical axis, the **amplitudes** and **propagation time** (in seconds or milliseconds) of the waves in the ground and, in horizontal axis, the **distance between the source and the geophones**. The amplitudes of the refracted waves are shown as a horizontal wiggle on the plot, or sometimes as colours. Generally-speaking, only the amplitudes of the compressional waves (P waves) are considered.

4.6.3 Output

The **first output** is a **time-distance graph** (figure 48*b*), which shows the travel time of the refracted waves as a function of the distance between the source and the geophones. Time-distance graphs are plotted by picking the travel times (also known as arrival times) of the refracted waves on the raw seismogram. The start of the wiggle should be taken for the picking of the travel time. The quality of seismic refraction measurements depends to a large extent on the care taken with the picking. The straight lines produced on the time distance graph by the picking are used to determine the velocities of the seismic waves in the stratified medium. At the x-axis origin, the line cuts the y-axis at a point called the intercept. The **seismic delay time** is, by definition, equal to half the intercept. For a stratified medium with n layers, and no dips, each delay time can be calculated using the equation:

$$D_{n-1} = \sum_{i=1}^{i=n-1} \frac{e_i \cos \theta_{c\,i,n}}{V_i}$$

where e_i is the thickness (m) of layer i, $\theta_{c\,i,n}$ the critical angle in layer i corresponding to the path to layer n and V_i is the velocity of the waves in layer i. The relationship derived from Snell's laws gives $V_i/V_n = \sin\theta_{c\,i,n}$.

Using this equation, the thickness below each geophone is determined and the **second output** is a **plot of the refractive boundaries as a function of depth and surface position.**

Other outputs, such as the interpretation of the time-distance graph using the **"plus-minus"** method may be used. This method is not described here, but is discussed in various references, such as Mari *et al.* (1998) or as Magnin et Bertrand (2005).

This method is used to explore the condition of the abrupt interface between the loose foundation of the dike and the *bedrock*. A zone where the homogeneity is disrupted in the body of the dike, or the presence of poorly-compacted zones in the loose foundation, may induce a velocity change that shows up on the time-distance graph, and which is characterised by **delays in the arrival times** (delay lags) corresponding to the refractive boundaries. **This anomaly will be found in all the time-distance graphs associated with the device**. Note that the delay must be significant compared with the error in measuring the travel time, *i.e.* about 3% greater than the time recorded.

a) **Principle**

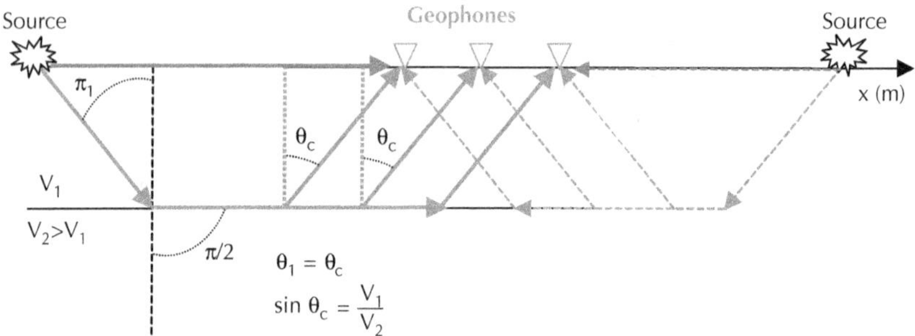

b) **Raw seismogram**

c) **Output: time-distance graph**

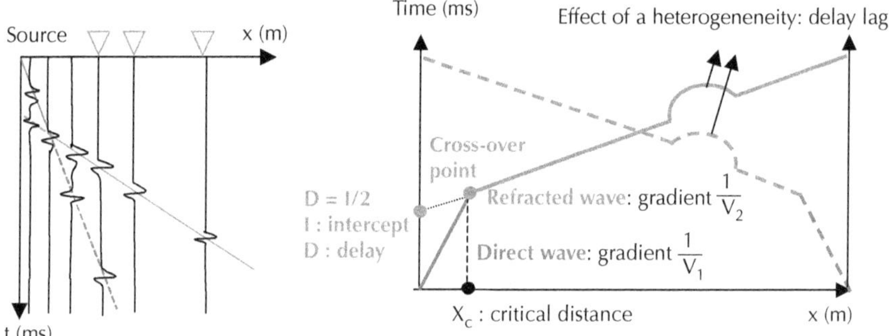

d) **Output: interpreted section**

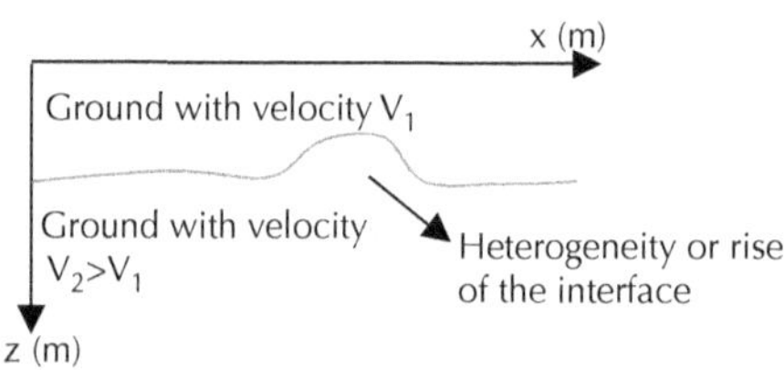

Figure 48 – *a)* Principle of the refraction of seismic waves showing direct offset and reverse shooting, *b)* qualitative illustration of a raw seismogram, *c)* time-distance graph and *d)* an interpreted section

Compared with the methods described previously, the strength of this method is its greater depth of penetration, essential for detecting bedrocks that often lie at depths of ten or more metres below the dike crest.

4.6.4 Methodology

The assumption of a profile of increasing velocity from layer to layer as a function of depth must first be confirmed. If this assumption is not found to be true, the method is not applicable. Prior modelling is a useful aid in interpreting the measured signals.

a) SELECTING THE DATA LOGGER

Seismic data loggers or seismographs record signals on a preset number of channels (classically 24 to 96 channels, but up to several hundred for some devices). These devices must offer the following features:

– a sampling interval of about 0.1 millisecond;

– a floating-point amplifier and an analogue/digital converter;

– a 50 and 60 Hz filter;

– the option for low pass, high pass and band pass filtering;

– the option to sum the shots (*stacks*);

– the option to save in standard format (SEG2).

b) SELECTING THE SEISMIC SOURCE

Depending of the required resolution, the source must be a high-energy source with a dominant frequency generally comprised between 10 Hz and 100 Hz. Ideally it should be light, be easy to handle, produce repeatable results and be safe. There are a number of possible sources, which fall into two categories: surface sources and buried sources (table 4).

Buried seismic sources produce more high-energy high frequencies, since they are in direct contact with deeper and more consolidated ground. For example, figure 49 compares two shots produced by different sources at the same site. The signals transmitted by the buried source generate higher frequencies (the time resolution is better), are less affected by noise and produce more energy (reflections between 30 and 40 ms are visible with a buried source, but invisible with a source on the surface). However, buried sources take longer to set up and damage the surface.

The high-frequency source is therefore selected to suit the surface conditions, the depth of the water table and the depth of the targets.

Surface source	
Impact	Hammer, weight drop
Projectile	Shotgun round (Betsy Seisgun)
Vibrator	MiniVib, MiniSosi
Buried source	
Explosive	Detonating cord, dynamite
Non-explosive	Blank shotgun cartridge, Sparker

Table 4 – Classification of high-resolution seismic sources (Bitri *et al.*, 1996)

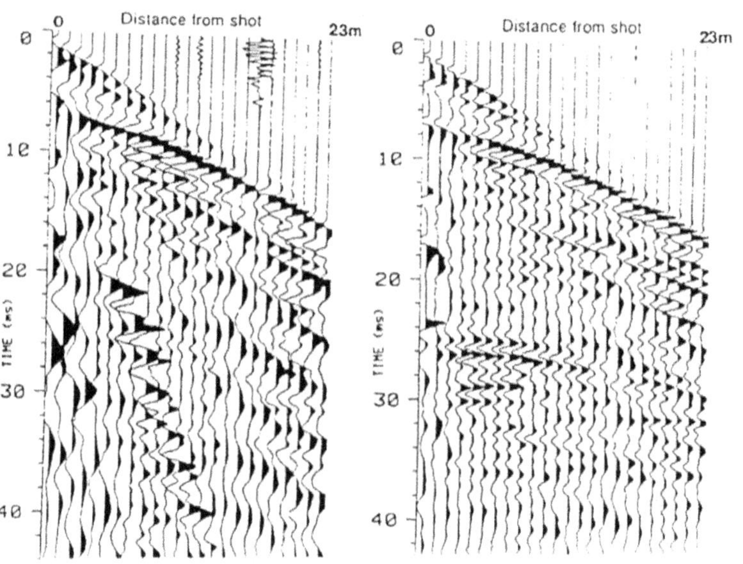

Figure 49 – Seismic sources: on the left, a shot produced by a surface source (hammer). On the right, a shot produced by a buried source (detonating cord) on the same site (Braham and McDonald, 1992)

c) Selecting the seismic receivers: geophones

Geophones are detectors that transform the energy of seismic waves into electrical energy. Geophones are often vertical, single-axis devices, since it is primarily the recording of the P waves that is of interest. They may be three-axis devices if there is a requirement to separate the response from P waves from that of S waves: in this case they measure the deformations in the three orthogonal directions on the surface.

The pulse response for a geophone is maximum at the dominant frequency, which is approximately 100 to 300 Hz for high-resolution seismic imaging. Figure 50 shows the behaviour of a 100-Hz geophone as a function of frequency. The correct operation of the geophone cannot be guaranteed at frequencies away from the frequency band recommended by the manufacturer. The bandwidth is between a few Hertz and a few hundred Hertz. Depending on the target being sought and the type of measurement, the dominant frequency of the geophones may be selected as follows (Bitri A. *et al.*, 1996): it must not be less than 10% of the maximum frequency that one is hoping to record.

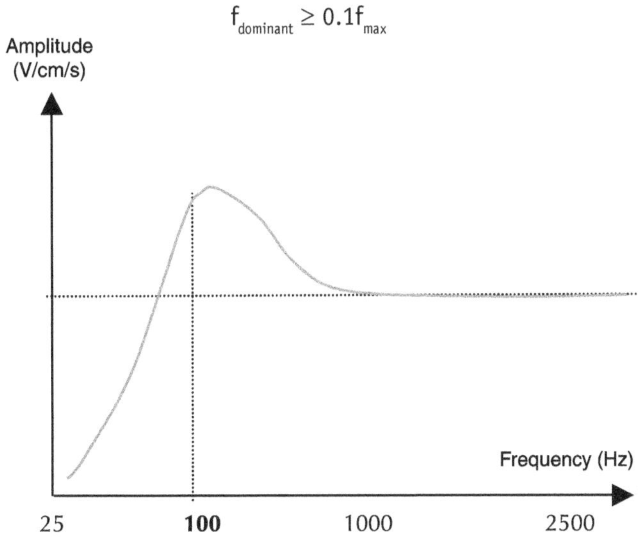

Figure 50 – Qualitative representation of the response of a 100-Hz geophone

d) Selecting the device size

The number of channels normally available on seismographs varies from 24 to 96. A high number of sensors minimises the need to move the device, and consequently provides greater measurement stability. The number of shots depends on the variability of the subsoil and on the accuracy required. For the case shown in figure 51, if the number of sensors n is 24, then 5 shots are fired – 2 end shots, 2 offset shots and one centre shot. Intermediate shots may be added between the end and centre shots if required and depending on the number of sensors used.

If the geological study, or an initial series of soundings, identified the first refractive boundaries, then the length L of the device will be chosen such that:

$$L \sim 2X_c$$

where X_c is the critical distance (from the shot to the cross-over point) (figure 51). In this case, the distance separating the offset shot (O or P) from the nearest geophone is about half the length of the device.

If no information is known about the subsoil, the operator sets the device length based on his or her own personal experience. A rough approximation of the investigation depth is about 1/6th of the device length. The distance separating the offset shot from the nearest geophone will, in this case, be about the same as the device length.

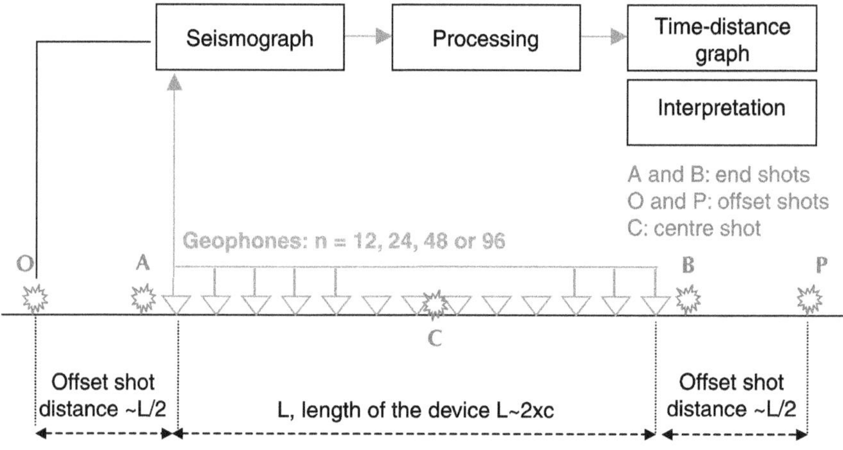

Figure 51 – Data acquisition system for seismic refraction

4.6.5 Example of measurements

The following example is taken from geophysical measurements made on the Canal du Centre (France) (Bièvre and Norgeot, 2004). Two profiles, each about one hundred metres long and determined using a device consisting of 24 geophones, 2-m apart, are shown in figure 53. These profiles have two shared geophones to overlap the measurements. The shots were produced by a weight drop. The time-distance plots (shown as black lines) consist of two segments, and the gradient of each segment of the line is inversely proportional to the velocity of the seismic waves in the medium: the first segment, with a velocity in this medium of about 500 m/s, over-lays a *substratum* in which the speed is about 4,000 m/s, corresponding respectively (refer to the section in figure 52) to the body of the earth dike and to the Jurassic substratum. These raw results were inverted using a simulated annealing method, whose initial data were the first arrival times of the seismic waves, the position of the sources and geophones and the velocities determined from the time-distance graphs. The result obtained is an image that shows the velocity in the medium as a function of depth. The change in the medium is indicated by a rapid change in the velocities on the image obtained: the profile (in black at the bottom of figure 53) shows the interface between the body of the dike and the substratum. Note that the velocity values given at this level by the inversion differ greatly from the values of the actual transition (500 m/s to 4,000 m/s). The result does, however, clearly reflect the actual situation in the field (correlation with sounding S3). It indicates, particularly, a reduction in the thickness of the dike body in an easterly direction.

The presence of a fault identified on the geological map and by the soundings is also revealed by the presence of a step in the offset shot (the most easterly time-distance plot, in blue, at the top of figure 53).

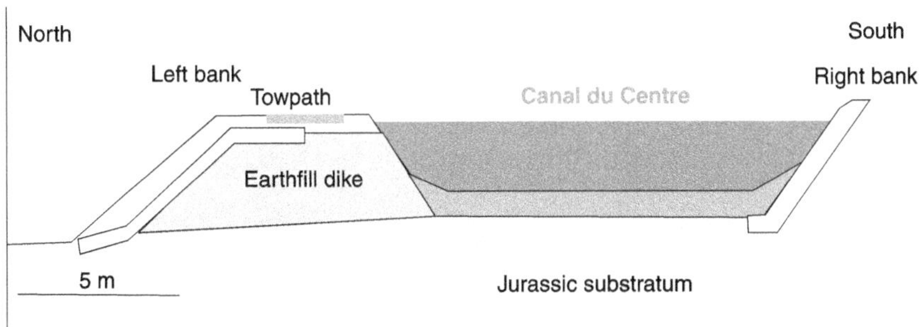

Figure 52 – Section across the Canal du Centre (CETMEF document, taken from Bièvre and Norgeot (2004)

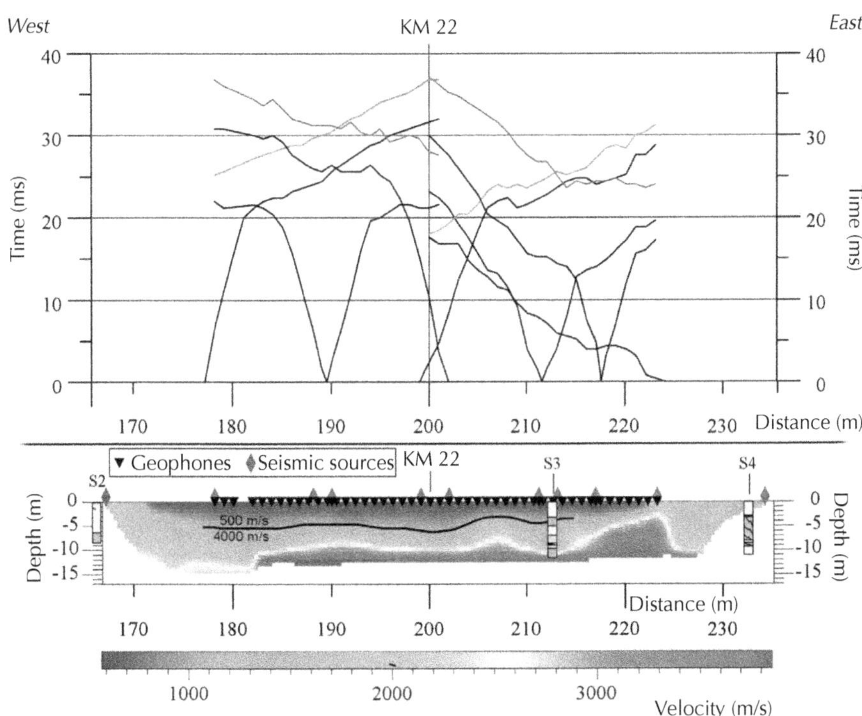

Figure 53 – Time-distance plots (at the top) and inversed measurements showing the velocities as a function of depth (CETMEF document, taken from Bièvre and Norgeot (2004)

4.6.6 Conclusion

Seismic refraction is only applicable to stratified media where the velocities in the layers increase with depth. It is useful in exploring or monitoring the condition of the contact between the loose foundation and the substratum. It is a local investigation method, whose rate of ground coverage is a few hundred metres per day. It generally requires two operators, one of whom should be an experienced geophysicist for interpreting the measurements.

The use of explosive sources and the associated noise (seismic and electromagnetic) are both limitations that restrict the use of the method in sensitive environments.

Effective geotechnical methods – In situ tests

5

5.1 Choice of methods and factors in locating investigation sites

As part of the "CriTerre" French project, a large number of geotechnical study reports was produced: these have been valuable in drafting this chapter, and include contributions from Wakselman (1999), Waschkowski (2001), Dupont (2000), Mieussens and Rojat (2002), as well as the methodological guide devised by Lino, Mériaux and Royet (2000). The methods described here are easy-to-use and effective, **the results of which are available *in situ*.**

Geotechnical test sites may be selected by a systematic investigation programme. However, and so long as this approach is agreed by the geophysicist and geotechnician, we advise basing the selection of the test sites on the results of preliminary studies and geophysical exploration. The geophysical method results are particularly important since they identify zones of heterogeneity that require more in-depth exploration by geotechnical means. By applying their experience, those interpreting the results can decide whether to extrapolate the results of local tests entirely or partially to the rest of the dike. They can also identify zones where local geophysical methods should be conducted (particularly transverse profiling by 2D electrical imaging). These zones may not be the same as those defined by the high-efficiency geophysical methods.

Checking geotechnical parameters on a local scale immediately provides a wealth of additional information for the diagnosis. It is also an essential step in interpreting the geophysical measurements.

5.2 Penetrometric tests: PANDA

5.2.1 Principle

PANDA is a French acronym for a computer-aided, dynamic, digital, autonomous, variable-energy penetrometer that has been developed by the civil engineering laboratory at Clermont-Ferrand University (Gourves, 1991). A standardised hammer is wielded manually to drive a tip into the ground at the end of a series of rods. The hammering energy is partially transmitted to the tip, which is moved deeper into the ground with each hammer blow, the distance travelled depending on the resistance of the ground.

For each impact, the equipment measures the penetration of the series of rods and the speed of the anvil.

5.2.2 Output

The output from the measurement is the penetration of the rod (in m) measured for each hammer blow, and the value of the resistance to dynamic penetration Q_d, calculated using the so-called "Dutch" formula (figure 54). The raw result is a penetration graph that plots Q_d as a function of penetration depth (in m). A smoothing is generally applied to the curves obtained.

5.2.3 Conditions for application

The Dutch equation is valid if:

– the ground exhibits perfect plastic behaviour during penetration;

– the lateral friction on the series of rods is negligible and all the energy supplied is transmitted to the tip, with the impact considered to be perfectly absorbed;

– the interstitial pressure in the ground is negligible.

5.2.4 Interpreting the results

Analysing the resistance to dynamic penetration as a function of depth provides information about the nature and density of the soils under consideration. This interpretation is made more precise if an accurate picture of the dike components is already known. It is therefore essential to consult the results of any preliminary studies and test drills. Relative variations in resistance may reveal the presence of voids. The dynamic resistance varies as a function of water content; knowing the piezometric level is important and the method should ideally be used on dry dikes.

Principle of the PANDA
A: cross-sectional area of tip (m^2)
e: plastic penetration (m)
M: striker mass (kg)
P: mass of the anvil of the series of rods (kg)
V: impact velocity (m/s)
Qd: resistance to dynamic penetration (Pa)

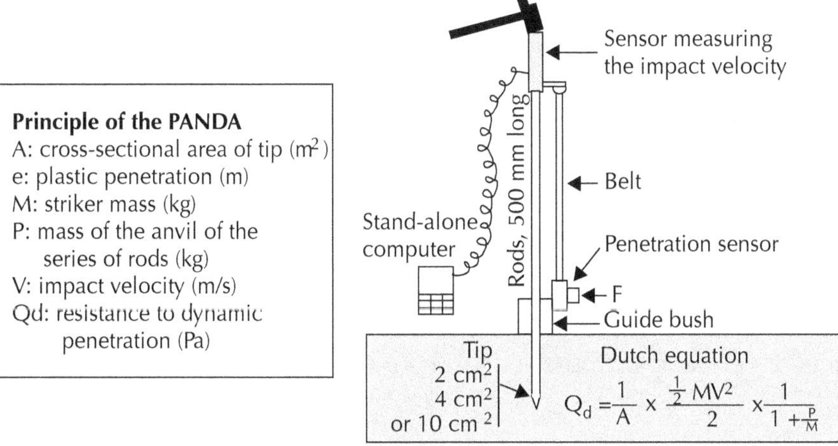

$$Q_d = \frac{1}{A} \times \frac{\frac{1}{2}MV^2}{2} \times \frac{1}{1 + \frac{P}{M}}$$

Figure 54 – Principle of the PANDA and the Dutch equation

5.2.5 Example of results

The following result was obtained on the dike on the Cher (figure 21). It is correlated with the results of a destructive sounding (SD 108 in figure 55). It reveals:

– a tarred road laid on an earth-fill zone with a total thickness of about 2.5 m;

– a dike body consisting of a layer of clay loam, 4 m thick;

– the foundation of the dike about 6.5 m deep and consisting of sands and gravel.

5.2.6 Strengths and weaknesses

The PANDA (Deplagne *et al.*, 1993) is, in principle, sensitive to variations in dynamic resistance that cannot necessarily be detected by test drills. As with all penetrometric tests, resistance peaks may be caused by the presence of localised heterogeneities, which are not necessarily representative of the medium as a whole. The investigation depth is between 5 and 10 m, depending on the materials. This lightweight and portable device (which makes it suitable for measuring sites with no vehicle access) is easy to use and accurate in identifying interfaces and for investigating the density of the dike body (for heights less than 10 m).

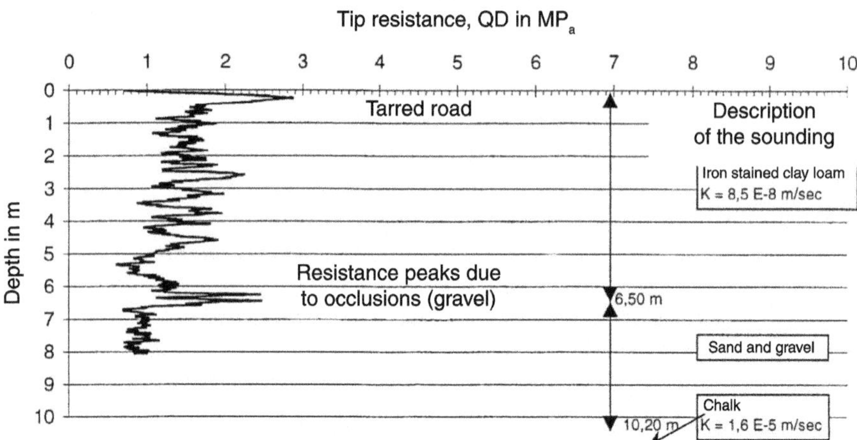

Figure 55 – Example of results obtained with the PANDA – Correlation with a destructive sounding (EDG document)

5.2.7 Characteristics of the equipment

– Length of each rod: 0.5 m. Diameter: 14 mm.

– Fixed tips, 2 cm^2 or disposable 4 or 10 cm^2.

– Water-proof data logger for recording resistance and penetration measurements.

– Total weight of about 20 kg.

– Storage capacity of 4,000 measurements, corresponding to one day of measurements.

– PC connection for data transfer and viewing results in situ.

– Two-person operation recommended, for speed and enhanced safety.

5.3 Penetrometric tests: Heavy-duty dynamic cone penetrometer (LCPC)

5.3.1 Principle

Testing with a heavy-duty, dynamic penetrometer (French standard NF.P.94-114) involves hammering an extended conical tip, with a basal area of 30 cm^2 and an apex angle of 90°, using a mechanical hammer weighing 35.5 or 96 kg, falling from a height of 75 cm. The number of blows required to achieve a penetration of 10 cm is recorded and used to calculate the unit dynamic resistance, q_d. To reduce the lateral friction on the rods, bentonite is injected behind the tip during hammering.

5.3.2 Result of the measurements

The output is the variation in dynamic resistance at the tip q_d (in MPa) as a function of tip depth (in m).

5.3.3 Interpreting the results

The interpretation is identical to that described for the PANDA.

5.3.4 Example of results

The results shown in figure 56 compare two tests, one using the PANDA and the other the heavy-duty penetrometer, conducted on the dike on the Cher river (figure 21). Overlaying the two penetration graphs reveals good correlation between the two methods; the advance increment for the PANDA is less (1 mm to 2 cm) than that for the heavy-duty penetrometer (10 cm). However, the heavy-duty penetrometer identifies the position of the interfaces more clearly.

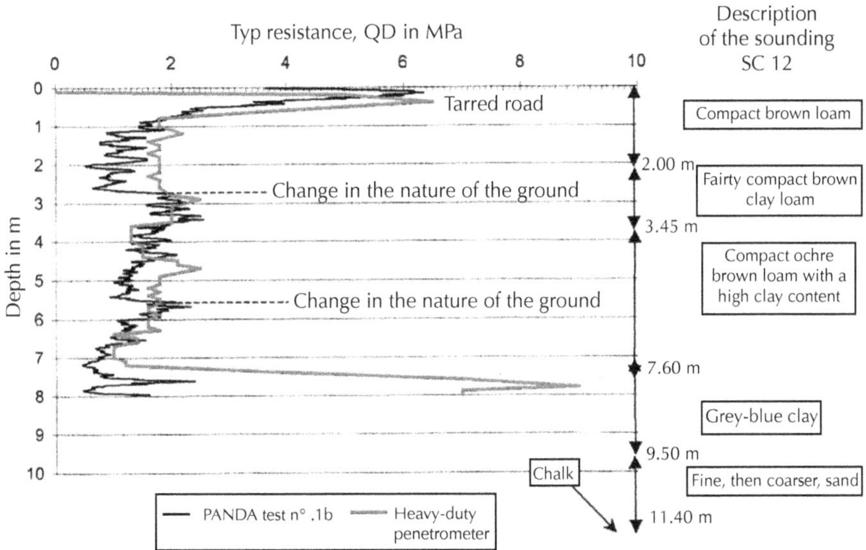

Figure 56 – Comparison of tests conducted with the PANDA and the heavy-duty penetrometer

5.3.5 Technical characteristics

– Standardised mass of the hammer: 64 kg.

– Other possible masses: 35.5; 96; 128 kg.

– Drop height: 0.75 m.

– Blow frequency: 20 to 30 blows/minute.

– Basal area of the tip: 30 cm^2.

– Length of one rod: 1 m.

– Mass of one rod: 4 kg.

– Injection head for a borehole test bed: dimensions: L = 4.50 m; l = 1.50 m.

5.3.6 Conclusions for penetrometric tests

Comparison of the PANDA with the heavy-duty penetrometer shows that the results of the methods are similar for the fine soil studied. The heavy-duty penetrometer indicates more clearly, but less accurately the interfaces between materials and offers a greater depth of investigation (> 10 m); unlike the PANDA device, it requires vehicle access to the measuring site. These tests are quick and easy to perform. The information collected is rapidly interpreted to give an indication of soil density and layer thickness.

5.4 Lefranc permeability tests

5.4.1 Principle

The Lefranc test (figure 57), carried out in association with borehole drilling, measures the permeability of the soil under investigation. Water is injected or pumped into an open-ended cylindrical cavity, called a lantern, at the bottom of the borehole and then the variations in hydraulic head and the corresponding flow rate are measured.

There are two types of test:

– **the constant head method:** this test is carried out in permeable soils (k > 10^{-4}m/s). The water is pumped and injected at a constant flow rate into the cavity until the level in the borehole stabilises. This test requires a significant amount of equipment (tank or pump and associated pipes and fittings). It is easy to use and the interpretation of the results is straightforward.

– **the variable head method**, normally carried out under the water table, in saturated soils: a given volume of water is injected into or taken from the lantern. Variations in the water level are monitored using a piezometric tube. This test is applicable primarily to poorly-permeable soils and does not require much equipment (20 to 60 litres of water). It is easy to perform, but its interpretation is more difficult.

The Lefranc test requires much more water if the soil is not saturated since the initial part of the test must saturate the ground. For dry dikes, the zones of interest are primarily located above the water table and, consequently, the constant head test should be preferred.

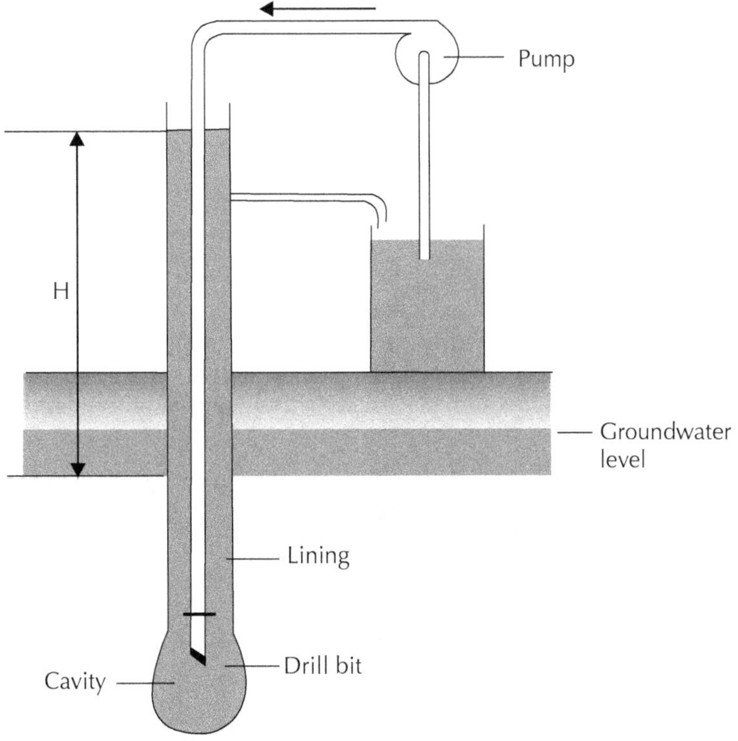

Figure 57 – Schematic of the Lefranc test

5.4.2 Quantity measured
The measured results give the local permeability around the lantern.

5.4.3 Conclusion
The Lefranc test is straightforward and inexpensive. It indicates the permeability of poorly-permeable soils. It does, however, need to be performed and interpreted by an experienced geotechnician.

5.5 Permeability tests using the Perméafor

5.5.1 Principle
This test is similar to the Lefranc injection test, during which the flow is controlled and the head is kept constant (Ursat, 1992 and 1995). It is carried out using:

– a drilling and water testing unit driven into the ground using a classical sounding machine operating in vibropercussion mode (hammering);

– a control and flow measurement module;

– a data logger and processing module.

The small interval between the measuring points means that a logging of the horizontal permeability can be obtained.

5.5.2 Output

The results are presented as the ratio of the injected flow rate Q (in $m^3.s^{-1}$) to the actual head H (in m) in the lantern.

The test duration (10 s) is not long enough to achieve a continuous flow in moderately or poorly-permeable soils. Therefore, the permeability values can only be interpreted for the most permeable soils (> 10^{-3} m.s^{-1}). Moreover, by recording the advance speed of the drilling tool the density of the soils can be evaluated.

5.5.3 Conditions for application

Ideally, the test should be used on moderately to highly permeable soils (10^{-3} to 5.10^{-3} m.s^{-1}). Quantitative interpretations in poorly-permeable soils are not recommended. The investigation depth is limited by the performance of the drilling equipment (typically 20 to 30 m). The rate of ground coverage offered by the test is approximately 30 to 60 m per day using the traditional procedure: one 10-s test every 20 cm.

5.5.4 Characteristics of the device

– Dimensions of the Permeafor module: 72*105*180 cm.

– Tests carried out as drilling progresses.

– Cylindrical perforated measuring drum, 5 cm high and 6 cm in diameter, *i.e.* to investigate a height within the 10-cm borehole. Bulges at the top and bottom of the perforated drum provide the sealing.

– Five to ten measurements per metre of borehole.

– Time for one standard test: 10 s.

– Quasi-horizontal flow into the soil through the perforated drum.

– Continuous calculation of the head losses in the equipment, used to take into consideration the actual pressure of the flow in the perforated drum.

– Immediate interpretation on site.

– Two operators: one controls the Permeafor and the other the drilling.

5.5.5 Conclusion

This test is quick and easy to use, requiring only classical drilling equipment. It measures, almost continuously, the horizontal permeability along the length of the borehole. Recording the advance speed also provides additional information about soil density. It does, however, require vehicle access to the measuring site. It is not suited to soils with low permeability, and the classical tests must be performed before a quantitative interpretation can be made.

5.6 Shear tests with the phicometer

5.6.1 Principle

The phicometer (Philipponat and Zerhouni, 1993) measures the shear strength of soils directly in a borehole. This mechanical property is obtained without needing to take undisturbed soil samples. This method is thus well suited to the characterisation of coarse or heterogeneous soils.

The phicometer is a probe that is inserted into the borehole, and consists of metal expansion shells fitted with horizontal annular teeth. These shells are rigid vertically, but can move laterally so that the teeth dig into the soil.

5.6.2 Quantity measured

The stress s, normal to the wall of the borehole, is kept constant using an inflatable cell. The series of rods are pulled until the soil fails in shear over the cylindrical surface S. The shear stress at failure is determined from this.

The test is repeated for increasing values of s. This gives several representative points in the Mohr's circle, which is used to plot the intrinsic straight line and to evaluate the friction angle j_i and the short-term residual cohesion C_i.

5.6.3 Conditions for application

The tests are carried out progressively as drilling work advances, by alternating the drilling and testing phases.

The quality of the drilling must be monitored.

Calibration must be performed so as to determine the shear surface area as a function of the probe volume.

Five to eight successive shear tests at constant s must be carried out; pulling at a constant speed of 2 mm/mn up to a 7 mm displacement per test.

The range of normal stress values should be determined as a function of the value (measured or estimated) of the limiting wall pressure P_l, as determined by the pressiometric test.

The method is not applicable to soft soils ($P_l < 0.3$ to 0.5 MPa).

Each test takes about one hour.

About four tests a day can be performed.

5.6.4 Characteristics of the device

– Phicometer probe connected to a pressure-volume controller.

– Series of rods that pass through a hollow actuator and a dynamometer at surface level that respectively apply and measure the pulling force.

– Probe diameter: 58 mm at rest, 80 mm when fully expanded.

– Useful length of the probe: 230 mm.

– Borehole diameter: 60 to 66 mm.

– Requirement for excellent borehole quality.

5.6.5 Strengths and weaknesses

The drilling quality must be excellent with no lining; which may be difficult to achieve in highly heterogeneous soils. This test is not suited to soft soils. It is one of the only tests that can determine the shear stress and the friction angle for soils *in situ,* and is very useful for some materials that are impossible to sample for laboratory tests (gravels, alluvium under the water table, flint clays, etc.).

5.7 Investigation using a mechanical shovel

This type of investigation involves digging a trial pit in the body of the dike or at its toe to check the distribution of the materials in situ. It is advisable to dig this trench with sloping sides so that the filling work conducted after the visual examination can compact the materials sufficiently.

This method lacks subtlety, but is straightforward and inexpensive, so long as there is adequate access to the site for a mechanical shovel. It is advisable to take reworked samples. Care should be taken to ensure that no-one climbs into the trench and that all relevant precautions are taken with respect to the risks of the trench walls caving in.

5.8 Mechanical drilling

Borehole drill testing is essential for the successful interpretation of geophysical measurements since it allows samples to be taken at specific locations. The devices are generally instrumented, and the **advance speed** is now at least one of the parameters that are almost systematically recorded during drilling. There are two types of drilling: core drilling and destructive drilling.

5.8.1 Core drilling

a) PRINCIPLE

Core drilling is a non-destructive means of taking soil samples. The method is time-consuming and fairly expensive, but provides an accurate picture of the distribution and mechanical properties of the materials in the dike. Moreover, subsequent laboratory tests can be conducted on these intact samples.

The core driller is driven into the ground on the end of a series of rods. The characteristics of the core driller are selected so that a complete identification and determination of the mechanical properties can subsequently be conducted in the soil mechanics laboratory. The diameter of the samples taken is generally between 75 and 100 mm.

b) RESULTS FROM A CORE DRILLING

The soil samples, or cores, are carefully identified, placed in wooden boxes, and then described and photographed by the geotechnician. The descriptions and information collected during drilling will be considered when interpreting the borehole logs.

The samples taken for laboratory testing are packaged immediately and carefully so as to preserve their water content (using paraffin) and structure (kept in a container). They are then labelled (reference, date and sampling method, top and bottom depths referenced to the surrounding ground), and then stored indoors protected from impacts, the sun and excessive temperature or humidity until taken away for laboratory testing.

c) CONDITIONS FOR APPLICATION

For fine, primarily sandy, soils the core sampler should be a ram-type drill with a stationary piston. It is sometimes impossible to take good quality samples from very sandy soils. For rocks or fine hard soils, rotary-type core drillers are used. These soundings must be performed in the presence of an experienced geotechnician.

5.8.2 Destructive drilling

a) PRINCIPLE

This type of drilling works by breaking up the materials using disintegrating drill bits and then transporting the fragments or cuttings to the surface using a circulating fluid or a helicoidal cutting tool. Drilling is performed:

– either with a mechanical earth auger, where the sampling tool is a cutting helix tool 100 to 300 mm in diameter;

– or with a down-the-hole drill or with a tricone bit (diameter of about 100 mm); the tool is driven into the ground by percussion or rotopercussion at the same time as the injection of a circulating fluid.

b) CONDITIONS FOR APPLICATION

Sounding with the earth auger is well suited to loose and poorly cohesive ground, whereas percussion sounding is suited to cohesive or rocky ground.

Sounding with the earth auger allows a simply and reasonably accurate determination of the nature of the materials. The samples are then subjected to identification tests (particle size, Atterberg limits, water content, etc.).

Interpretation of the percussion or rotopercussion soundings is more difficult, since it is based on the observation of *cuttings* that are transported to the surface at various and irregular times by the circulating fluid. The recording of the drilling parameters (advance speed, tool pressure, pressure of the circulation fluid, etc.) refines the interpretation.

Equally, the presence of a geotechnician and a carefully filled in observation book are recommended when using and interpreting the results of the drilling.

5.8.3 Strengths and weaknesses

Destructive or non-destructive drillings provide accurate information about the structure of the soil, and can be used to calibrate geophysical measurements. The borehole may be fitted with a piezometer or inclinometer type sounding instrument. Finally, the intact samples produced by core drilling are valuable since they can be sent off for more sophisticated analysis in the laboratory.

The cost remains relatively high for information about a single location. These methods require access and sufficient space on the ground to deploy the drilling vehicle. Finally, the results of the tests on core samples only provide single-point information about the composition of the body of the dike or its foundation.

6

METHODOLOGICAL SUMMARY OF THE GEOPHYSICAL AND GEOTECHNICAL TECHNIQUES USED FOR THE EFFICIENT DIAGNOSIS OF DRY DIKES

6.1 Reminder of the context of the study

Dikes are very long structures (several kilometres to several dozen kilometres long) the height of which ranges from 3 to 12 m, depending on the nature of the dike and the flood levels that it is designed to contain.

A successful diagnosis of this type of structure would determine the degree of safety provided by the dike and would identify all the points of weakness and anomalies[4]. The findings could then be used either to define the remedial work required or as a basis for a more in-depth diagnosis.

The starting point for the study is to consider all the components of the dike and all the systems with which it interacts. The methods used, ideally high-efficiency methods, must identify and assess, as accurately as

____

4. Note that a single point of weakness can compromise the effectiveness of the entire dike protection system, as was clearly demonstrated in France by the disastrous flooding of the Rhone in December 2003.

possible, all the features of the structure that might suffer irreversible damage with major consequences (particularly overtopping and breaches), in the event of the river flooding. To attain this objective, the following methodology is proposed.

6.2 Key study phases in the diagnosis

The general methodology for the diagnosis is shown in figure 58. It consists of three key phases:

– preliminary studies;

– geophysical exploration;

– geotechnical exploration.

There are two possible study pathways, depending on whether local exploration using geophysical methods is performed after the high-efficiency geophysical methods (blue pathway in figure 58), or alternatively after the geotechnical tests (orange pathway in figure 58).

6.2.1 Preliminary studies

The diagnosis necessarily starts by conducting **historical research** into the structure. This work is performed in conjunction with a **geological study**, a **morphodynamic analysis** of the watercourse, a **topographical survey** along and across the dike and a **visual inspection** of its entire length.

This work is time-consuming and requires multi-disciplinary skills and a variety of working methods. However, they are an essential prerequisite to interpreting the measurements made during subsequent phases.

6.2.2 Geophysical methods

The next phase in the diagnosis is **high-efficiency geophysical methods**. These measure a physical quantity that provides information about the distribution of the materials in the body of the dike and down to its foundation. The purpose of these methods is to identify portions of the dike, generally on a scale of tens of metres, that reveal changes in the composition of the structure, or that highlight specific features cutting across the body of the dike, such as conduits or pipes.

At the present time, the methods best suited to satisfying the requirements of high efficiency (*i.e.* collecting a large amount of good quality data over long distances) are **Slingram-type low-frequency electromagnetic methods** (with two coils), that provide information about the distribution of materials through measuring the **apparent resistivity** ρ_a **(Ω.m)** – the inverse of the apparent conductivity that relates to the electrical conductivity of the soils. If these methods cannot be applied, for site-specific reasons, **radio magnetotellurics (RMT)** can be used, with due consideration of the reservations relating to this method expressed above. If the dike is low (< 5 m) and the materials are relatively resistant (> 100 Ω.m), then ground penetrating radar can be used.

It is very good practice to apply **local geophysical methods** to specific zones identified by high efficiency geophysical methods. Although the rate of ground coverage is much less, the results are more precise, giving a more accurate picture of the distribution of the materials in such zones. This phase may be carried out after the high-efficiency methods, or alternatively after the geotechnical tests.

For dry dikes, the experiments conducted in the context of the "CriTerre" French national Project and in other contexts have highlighted the wealth of information provided by apparent resistivity measurements made by **electrical imaging surveying (2D electrode array)**. This method is particularly effective in producing a transverse or longitudinal tomography of the structure.

Detecting the interface between the body of the dike and its foundation is another valuable element in establishing a diagnosis: **seismic refraction** is well suited to this task since it monitors the contacts between layers, so long as the velocity of the seismic waves in the medium increases as a function of the depth.

For all the methods, an experienced geophysicist should be employed to perform the interpretation (2003-2004). Such methods are still under investigation within the ERINOH ANR project.

One way of guaranteeing the high quality of the implementation and interpretation of geophysical exploration work carried out on a site is to stipulate, in the definition of the requirements for the diagnostic studies, that a **test of the repeatability** of the method(s) be carried out on a carefully-chosen test section of the dike. This preliminary test should be performed before the geophysical investigation itself, coordinated by and paid for by the studies office responsible for the overall investigation, and the results should be interpreted and commented on by an experienced geophysicist. A contractual arrangement of this kind can only reasonably be envisaged for methods offering high efficiency (Slingram, RMT, GPR, etc.).

At this stage in the diagnosis, the results of the preliminary studies correlated with the results of geophysical studies should allow an initial interpretation of the dike in general terms. However, this is not the comprehensive and high-quality diagnosis that is required. It is therefore essential to carry out the geotechnical study phase.

6.2.3 Geotechnical methods

Effective geotechnical methods are *in situ* methods that can confirm certain characteristics of the materials. **Penetrometric tests** are systematically applied to dikes. They very quickly reveal the stratification of the structure, and give a qualitative approximation of the mechanical properties of the materials.

The **Lefranc and Permeafor methods** measure the horizontal permeability of the soil at various points down a borehole. These tests are easy to perform and inexpensive. Recording other parameters such as the speed of advance of the drill bit provides additional information that helps in interpreting the measurements in relation to soil density.

Core drilling is more expensive but provides a great deal of information. This techniques removes intact "cores" or soil samples for subsequent testing in the soil mechanics laboratory where the mechanical properties, nature and grain size of the materials can be analysed. **Destructive drilling**, less expensive than core drillings, is a less sophisticated *in situ testing method and can also be used for the insertion of* surveying instruments (*e.g.* piezometers).

Most drilling equipment is now instrumented, and one commonly-recorded parameter is the advance speed of the bit; this provides information immediately about the stratification and mechanical properties of the investigated soil.

Finally, the **phicometer test** measures, in situ in the borehole, the shear strength of the materials, and is especially useful for testing those materials that cannot be extracted as cores.

For all the tests, it is highly advisable to employ an experienced geotechnician to assist with the testing and to interpret the results.

6.3 Reporting the geophysical and geotechnical results

A description of how the results of the preliminary studies are reported is available in two methodological guides (Mériaux *et al.*, 2004[5]; Lino *et al.*, 2000).

The overview of a geophysical and geotechnical study generally includes:

– a presentation of the context and of the study site (cross section of the dike and longitudinal profile drawing), whose position is located using a reference system that is applicable to all the phases of the diagnosis (preliminary, geophysical and geotechnical studies);

– a summary of the observations made during the preliminary studies (visual inspection, historical indicators and geological study). For the geophysicist to conduct his or her phase of the work satisfactorily, as much information as possible should be provided about the composition and features of the dike. If the preliminary studies phase is omitted, the geophysical methods have to be deployed "blindly" and the results will be incomplete;

– a summary description of the methods used;

– an efficient presentation of the measurements (profiles, maps, logs, tables) with their position located using the reference system set initially and using standard units. The results of the geophysical measurements must be commented on and correlated with the information available. The results of the geotechnical investigations are initially presented independently and generally compared with the assumed or known cross section of the structure. It is then advisable to consider the results in the light of the results of the geophysical exploration, so as to show the correlations between the measured quantities;

---

5. Soon available in English (Mériaux, Royet 2007).

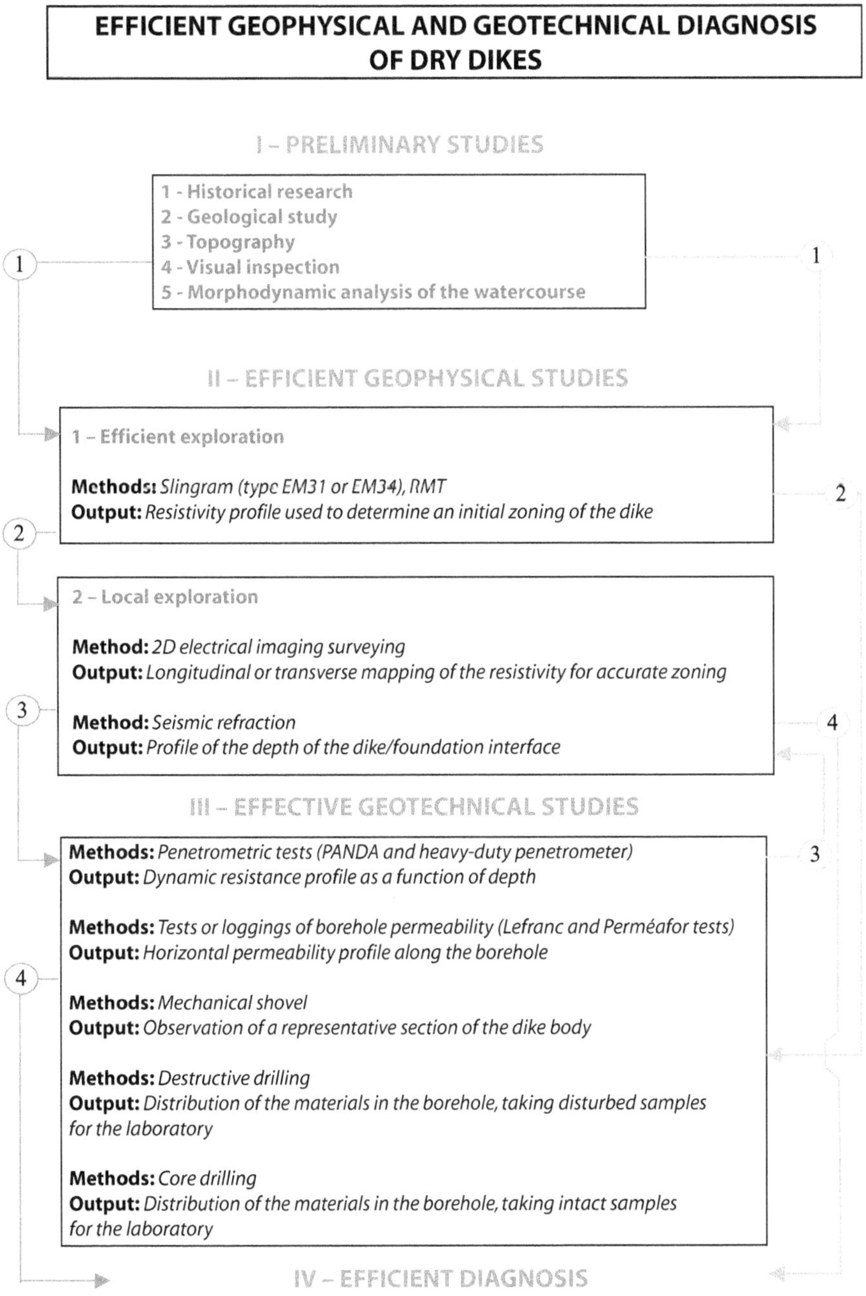

Figure 58 – General methodology for dike diagnosis and possible study pathways. In blue, local geophysical methods are conducted after the high-efficiency methods; in orange, after the geotechnical tests mentioned

– a conclusion summarising the high-efficiency study. This must assist the directors of works in deciding on the work to undertake or in defining further, more in-depth studies on certain portions. Laboratory tests and modelling studies may then be envisaged.

To guarantee the quality of the services provided, the results of the geophysical and geotechnical methods must be interpreted, respectively, by an experienced geophysicist and geotechnician.

6.4 Comparison of the geophysical methods

Based on the experiments conducted in the context of the CriTerre national project and on other laboratory reports and design office studies, an evaluation of the geophysical methods described in this guide is given in figure 59. They are all suitable to the efficient diagnosis of dry dikes. The efficacy of the methods has been attributed based on their ability to explore the dike body and the upper part of its foundation and to locate the heterogeneities likely to be present in the dike body (distribution of conductive materials, pipes and conduits, internal flows of water, etc.).

The **Slingram electromagnetic method** is the method of choice for dike diagnosis since it is suited to investigating the range of heights typical to this type of structure. It offers high efficiency for dikes whose crest can be driven on, and can explore down to depths of about 6 m. For greater depths (10 to 15 m), the rate of ground coverage is about the same as the walking pace of the operator(s). The variable measured is the apparent resistivity. This data gives an overall picture of the distribution of materials in a dike volume defined by the depth of investigation of the method.

Method	Effectiveness	On dikes suited to vehicules	On dikes unsuited to vehicules
Slingram	+++	yes	yes, if portable device
RMT	++	yes, if capacitive electrodes	yes, if conductive electrodes
2D electrical imaging	+++	difficult on a tarred road	yes
2D capacitive electrical imaging	+	yes	no
Seismic refraction	++	yes	yes
Ground penetrating radar	+	yes	yes

Figure 59 – Efficacy of geophysical methods for exploring dikes

For those cases where the Slingram method is not applicable on the crest, **RMT** can be considered if the dike crest can be driven along. The raw results of this method should be interpreted with care and the stability of current devices is not completely reliable. On dikes, at least three runs along the same profile are required to obtain a usable measurement.

Ground penetrating radar is not applicable to the majority of the body of the dike. However, this method is suited to certain configurations (low, dry dike consisting of sufficiently resistant material).

Finally, **2D electrical imaging surveying** is not a high-efficiency method, but does give an accurate picture of the distribution of materials in the dike body. It is a valuable complement to the high-efficiency methods, particularly in terms of providing a means of calibrating the geophysical measurements to the local context.

The main characteristics of these methods are shown in figure 60:

	Depth of investigation	Rate of ground coverage	Type of target detected by the method	Sensitivity to noise	Conditions required for detection
Slingram	Theoretically down to 50 m	One to two km/day for portable devices, several km to several dozen km/day for towable devices	Resistant zones and buried pipes and conduits	Electric power lines and electric fences	Conductive medium
RMT		Low if conductive electrodes are inserted in soil, several km to a few dozen km/day for towable devices		Dike topology, electric power lines, metal fences, uneven crest surface	Reception from transmitters, dike must not have high sinuosity
Seismic refraction	Down to 30 m	A few hundred metres per day for devices one hundred metres long and with an inter-geophone spacing of 1 or 2 m	Interface between layers with velocity increasing as a function of depth	Traffic, micro seismic events	Velocity in the medium must increase with depth
GPR	Inversely proportional to conductivity. Experimentally for dikes built with resistant materials, 2 to 5 m on the crest, about a dozen m at the foot	Several dozen to several hundred km per day	Pavement system, sandy-gravelly parts of dike	Radio transmitter, metal elements nearby	The resistivity of the medium must be greater than 100 Ω.m
2D electrical imaging	Depends on the line length: 1*R for a free space sphere of radius R and 1.5*R for a conductive sphere of radius R	A few hundred metres per day, for devices one hundred metres long and with 1 or 2 m between electrodes	Anomalies ideally conductive (dips, stratification)	Ground less conductive than the target	/

Figure 60 – Main characteristics of geophysical methods for the high-efficiency exploration of dry dikes

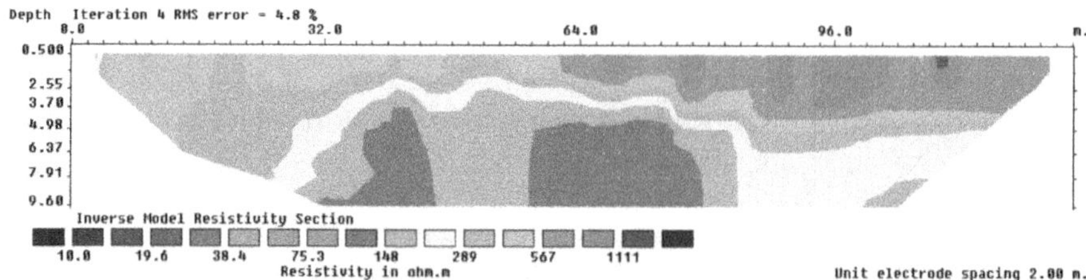

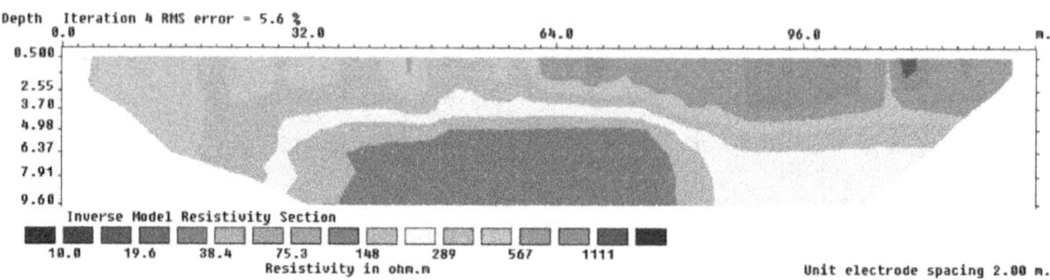

Figure 61 – Repeatability test of 2D electrical surveys on a 100 m section – dike right bank of the Romanche river at Vizille (French Alps) – from IMS doc

Figure 62 – Exploration using the FAST-EM device – developed by the French company SOBESOL – on a canal embankment near Sarreguemines (Moselle, France)

PROPOSAL FOR THE DEVELOPMENT OR TESTING OF GEOPHYSICAL AND GEOTECHNICAL METHODS FOR DIKE DIAGNOSIS

In France, geophysical methods have been applied to the diagnostic investigation of dry dikes since the end of the 1990's. These methods, initially developed for subsurface exploration, may open up new avenues of investigation and research.

Slingram electromagnetic methods, which are currently the first choice method for the high-efficiency investigation of dry dikes, may be further improved for this application. Indeed, the devices used to carry out high-efficiency investigations that are currently available on the market offer only a limited depth of investigation (about 6 m). To reach greater depths, the intercoil spacing must be increased – to 10 m to explore a depth of 10 m. This configuration requires two operators on foot. The rate of ground coverage is then low, and the studies will inevitably be long and laborious when diagnosing long dikes. The time would appear to be right for the development of an adapted Slingram device that can be towed behind a vehicle.[6]

---

6. The French company SOBESOL has developed the FAST-EM® system – a Slingram device that is towed behind a light off-road vehicle.

Even though its first practical application was in the context of "CriTerre" experiments, the RMT method, whilst performing well in deep investigation applications such as earthworking, suffers from a number of shortcomings when used on dikes. The topographical effects of the dike on the field measured (sinuosity and gradient of the slopes in particular) have not yet been mastered. The number of transmitters available at nearby frequencies and whose fields are orthogonal at the point of measurement is insufficient. Finally, and this is a major drawback, the use of capacitive mats and the sensitivity of the measurements to nearby surface heterogeneities significantly reduces the efficiency initially expected from this method and impairs the quality of the interpretation. If this method is to be used on dikes in the future then specific developments will be required, and whilst the focus of this development is currently unclear, it could relate particularly to the type of electric field sensor used. Moreover, investigation of "inclination" type methods that do not require contact with the ground could provide a solution.

Finally, other seismic methods have been suggested, particularly high-resolution seismic reflection and surface wave seismic techniques. The first could identify the various reflective boundaries local to the device, and its application is not restricted to subsurface layers in which the velocity increases with depth. The second method is a device that is fitted with sliders, thus allowing more high-efficiency applications. For these two methods, the potential value of the results would appear to be worth studying with a few tests.

Effective geotechnical methods, *i.e.* the methods used to test materials in situ, may also benefit from the wealth of information provided by the logging of natural radioactivity (gamma-ray log). This technique measures the natural radioactivity of the ground being sounded. The principle is based on the emission of gamma photons by the radioactive decay of natural elements such as uranium, thorium and potassium and on the reception of these particles by a probe placed in a borehole. In sedimentary media, it is useful in revealing clayey layers, which are highly radioactive in their natural state. The technique is also used to define the stratification of the ground locally when it is carried out in a single borehole, and to plot the formation of the soil when probes are placed in several boreholes. This method is very simple to use and does not pose any radiation risks.

These topics are the subject of open discussion. In the future, research work applied to geophysical and geotechnical methods will very probably lead to the emergence of new methods that provide better solutions for the directors of works. The methodology proposed in this guide and the description of the methods has taken into consideration information fed back from research conducted in the context of the "CriTerre" French national Project and on several recent investigation sites. Methodologies and methods will certainly need to change in the years to come to keep pace with increasing demand diagnostic work and to integrate more sophisticated techniques.

8 CONCLUSION

This guide draws on the latest findings and information available to present a general methodology for diagnosing dry dikes, and particularly the methodology recommended for implementing and interpreting adapted geophysical and geotechnical methods. The guide begins by setting the context of the study and giving a detailed description of the main characteristics of river dikes.

It continues with a brief description of the preliminary studies required for a successful diagnosis. These studies include historical research, geological surveys, visual inspection of the structure, a topographical study and morphodynamic analysis of the river. The part on geophysical exploration methods describes the physical quantities that are measured to determine, in approximate terms, the distribution of the materials in the body of the dike. The principal variable measured is the apparent resistivity – the inverse of the apparent conductivity – that is associated with electrical conductivity phenomena in soils, and which is processed to produce an "image" of the dike body and of some of its foundations. High-efficiency low-frequency electromagnetic methods such as the Slingram method (twin coils) are the method of choice for efficient diagnosis.

The radio magnetotelluric (RMT) method is presented for historical reasons. Various experiments and numerous experts have highlighted the difficulties inherent to its implementation and interpretation.

The results of these high-efficiency methods must be considered in the light of the findings produced by local methods that provide a more accurate picture of the distribution of the materials actually present. One valuable local method is electrical imaging surveying (2D electrode array), which produces a transverse or longitudinal image, in general terms, of the geometry of the dike. Another technique is seismic refraction, which explores the interfaces between the body of the structure and its foundation. Other methods are described and may be considered if they are suited to the specifics of the site. One such example is ground penetrating radar, although there are certain limitations associated with this technique, as described in the text.

The interpretation of the geophysical measurements (phase 2) must be correlated both with the preliminary studies (phase 1) and with the geotechnical tests (phase 3). This correlation is an essential condition for a high-quality diagnosis. Geotechnical tests are described briefly. These tests are effective in that they produce valuable information *rapidly and in situ* about the component materials in place in the structure: testing with a dynamic penetrometer, permeability tests, sampling with a mechanical shovel, destructive drilling and core drilling.

So long as these three phases (preliminary studies, geophysical measurements and geotechnical tests) are conducted correctly, the efficient diagnosis of a dry dike can be comprehensively made. The various studies consulted when putting together this guide revealed that, at each phase, the involvement of experienced professionals (geophysicists, geotechnicians and specialists in hydraulic structures) is an essential element in producing a high-quality study.

The aim of this guide is to describe all the methodological considerations involved in establishing a diagnosis of a dry dike. It was written after consulting numerous studies conducted by private design offices and public institutions, and after consulting several experts in the field of geotechnics, applied geophysics and hydraulic structures, with a view to putting forward a consensus regarding the methods proposed. It does not aspire to cover all aspects of these broad fields, but rather is intended to provide a tool to assist in the decision-making process for directors of works and engineers responsible for carrying out diagnostic studies on "dry" dikes[7].

---

7. That is: flood protection dikes not subjected to hydraulic head at the time of the exploration.

BIBLIOGRAPHY

● ● ● **Reports and papers written within the context of the "CriTerre" French national Research Project (1999-2004)**

Cher experimental site:

Dupont L., 2000. *Diagnostic de l'état des digues de la Loire : reconnaissance des anomalies géotechniques (dossier n° CRT/00/1/005)*. CETE de Blois/Projet national CriTerre.

Mériaux P., Royet P., Côte P., Hollier-Larousse A., Frappin P., 2001. *Méthodes de reconnaissance géophysique à grand rendement pour les digues de protection contre les crues*. Colloque GEOFCAN, septembre 2001, Orléans, 111-115.

Waschkowski P., 2001. *Rapport de sondages*. LRPC de Blois.

Wakselman J., 1999. *Étude comparative des méthodes de reconnaissance géophysique et géotechnique des digues à sec : projet CriTerre d'expérimentation sur les levées de la Loire*. Mémoire DESS de géophysique appliquée, université Pierre et Marie Curie/ Cemagref.

Cher experimental site: reports consulted but not mentioned in the guide

Dondaine E., 1999. *Projet CriTerre 1.3 – Méthodes de reconnaissance à grand rendement – application aux digues, levée du Cher au confluent avec la Loire – Prospection par panneau, traîné électrique et sismique réfraction.* Rapport EDG.

Hollier-Larousse A., 2000. *Reconnaissance géophysique par radio-magnétotellurique : projet national CriTerre, thème anomalies physiques, reconnaissance à grand rendement.* LCPC Nantes/Projet national Criterre.

Hollier-Larousse A., Folton C., Mériaux P., Wakselman J., Frappin P., Côte P., Royet P., 2000. *Étude comparative des méthodes de reconnaissance géophysique et géotechnique des digues à sec – site d'expérimentation de la digue du Cher à Savonnières.* Rapport général de recherches, LCPC de Nantes/Cemagref/EDG/Projet national CriTerre.

Hollier-Larousse A., Mériaux P., Waschkowski P., Mieussens C., 2002. *Les hétérogénéités des digues de protection contre les inondations : leur signature géophysique et géotechnique.* Journées AGAP 2002.

Waschkowski P, 2000. *Détection des anomalies et contrôle des améliorations des sols – Diagnostic de l'état des digues de Loire.* Rapport technique, projet national Criterre, LRPC Blois, juillet.

Agly experimental site

Frappin P., 2001. *Étude géophysique dans le secteur de l'Agly – prospection géophysique [EM31, panneaux électriques].* EDG/Projet national CriTerre.

Hollier-Larousse A., 2001a. *Reconnaissance géophysique par radio-magnétotellurique : Digue [rive gauche] de l'Agly à Saint-Laurent-de-La-Salanque 2000-2001 – Projet national CriTerre, thème : anomalies physiques, reconnaissance à grand rendement.* LCPC Nantes/Projet national CriTerre.

Hollier-Larousse A., 2001b. *Reconnaissance géophysique par radio-magnétotellurique : Digue [rive droite] de l'Agly à Pia 2000-2001 – Projet national CriTerre, thème : anomalies physiques, reconnaissance à grand rendement.* LCPC Nantes/Projet national CriTerre.

Mieussens C., Rojat F., 2002. *Reconnaissance des digues de l'Agly – Essais au pénétromètre dynamique. Communes de Pia et de Saint-Laurent-de-la-Salanque.* LRPC de Toulouse.

Mériaux P., Hollier-Larousse A., Dérobert X., 2003. *Reconnaissance géophysique des digues de l'Agly après la crue de novembre 1999 – Contribution à l'élaboration d'une méthode de diagnostic.* Colloque GEOFCAN, septembre 2003, Paris.

Fauchard C., Mériaux P., 2004. *Les méthodes de reconnaissance à grand rendement adaptées aux digues de protection contre les inondations.* Colloque CFGB, 25 et 26 novembre 2004, Orléans.

Agly experimental site: reports consulted but not mentioned in the guide

Dérobert X., 2001. *Agly : Radar géophysique [commune de Saint-Laurent-de-la-Salanque].* LCPC de Nantes.

Fauchard C., Mériaux P., Durand O., 2004. *Répétitivité de la méthode Radio-MT sur les digues de l'Agly. Thème anomalie physique – Reconnaissance à grand rendement.* LCPC Nantes, Cemagref.

Hollier-Larousse A., Mériaux, P., Waschkowski, P., Mieussens, C., 2002. *Les hétérogénéités des digues de protection contre les inondations : leur signature géophysique et géotechnique.* Journées AGAP 2002.

Hollier-Larousse A., Folton C., Mériaux P., Frappin P., Luquet C., 2004. *Méthodes géophysiques et géotechniques à grand rendement pour la reconnaissance des digues à sec. Site d'expérimentation des digues de l'Agly.* Rapport général de recherches, LCPC de Nantes/Cemagref/EDG/Projet national CriTerre.

Articles, guides and other reports

AGAP-collectif, 1992. *Code de bonnes pratiques en géophysique appliquée – 75 fiches et guide d'application.* Édition AGAP Qualité.

Bièvre G. et Norgeot C., 2004. Utilisation des méthodes géophysiques pour l'auscultation des digues : étude de cas sur le canal du Centre (France). *Bulletin des laboratoires des Ponts et Chaussées,* à paraître.

Bitri A., Perrin J., Beauce A., 1996. *Sismique réflexion haute résolution : principe et applications.* Rapport du Bureau de recherche géologique et minière R 39220, 59 pages.

Bosch F.P. et Müller I., 2001. Continuous gradient VLF measurements : a new possibility for high resolution mapping of karst structures. *First Break,* 10 (6), 343-350.

Bourgeois B., Suignard K. et Perrusson G., 2000. Electric and magnetic dipoles for geometric interpretation of three-component electromagnetic data in geophysics. *Inverse Problems,* vol. 16, 1225-1261.

Braham P.J. McDonald R.J., 1992. Imaging a buried river channel in an intertidal area of South Wales using high resolution seismic techniques, *Quaterly Journal Engineering Geology,* vol. 25, 227-238.

Cagniard L., 1953. Principe de la magnéto-tellurique, nouvelle méthode de la prospection géophysique, *Annales de géophysique,* tome 9, fascicule 2.

Chevassu G., Grisoni J.-C., Lagabrielle R., 1990. "Utilisation de la magnétotellurique pour l'auscultation du sol-support de chaussée, *Bulletin de liaison des laboratoires des Ponts et chaussées,* n° 166, 73-82.

Chouteau M., 2001. *Méthodes électriques, électromagnétiques et sismiques,* Géophysique appliquée II, GLQ 3202, Notes de cours, École polytechnique de Montréal.

Combes P.-F., 1996. *Mico-ondes 1. Lignes, guides et cavités.* Paris, Dunod.

Daniels D.-J., Gunton D.-J., Scott H.-F., 1988. Introduction to subsurface radar, *IEE Proceedings,* vol. 135, Pt. F, n° 4.

Daniels D.-J., 1996. Surface Penetrating Radar, *IEE Radar, Sonar, Navigation and Avionics,* Series 6.

Davis J.-L., Annan A.P., 1989. Ground-penetrating radar for high resolution mapping of soil and rock stratigraphy, *Geophysical Prospecting,* 37, 531-551.

Degoutte G., 1992. *Guide pour le diagnostic rapide des barrages anciens,* Paris, Cemagref.

Degoutte G. coord., 1997. *Petits barrages, recommandations pour la conception, la réalisation et le suivi,* Paris, CFGB/Cemagref Éditions.

Deplagne F., Bacconnet C., Royet P., 1993. *Intérêt du pénétromètre léger pour le contrôle du compactage des barrages en terre.* Journées nationales d'études "Petits barrages", AFEID-CFGB – 2 et 3 février 1993.

Eyraud L., Grange G. et Ohanessian H., 1973. *Théorie et technique des antennes,* Paris, Librairie Vuibert.

Gourves R., 1991. *Le PANDA. Pénétromètre dynamique léger à énergie variable pour la reconnaissance des sols,* Laboratoire de génie civil. CUST. université Blaise-Pascal.

Guérin R., Tabbagh A., Benderitter Y., Andrieux A., 1994. Invariants for correcting field polarisation effect in MT-VLF resitivity mapping, *Journal of Applied Geophysics,* vol. 32, 375-383.

Halbecq W., 1996. *Approche géomorphologique des brèches dans les levées de la Loire.* Mémoire DEA de l'université d'Orléans.

Hollier-Larousse A., 1997. *Contribution à la valorisation d'une méthode géophysique électromagnétique utilisée en géophysique appliquée de subsurface : la radio magnéto-tellurique,* Rapport du Conservatoire national des arts et métiers, Paris.

Lagabrielle R., 1986. *Nouvelles applications de méthodes géophysiques à la reconnaissance en génie civil,* Rapport des laboratoires des Ponts et chaussées.

Lino M., Mériaux P., Royet P., 2000. *Méthodologie de diagnostic des digues appliquées aux levées de la Loire moyenne.* Paris, Cemagref Éditions.

Loke M.H., 1999-2002. *Electrical imaging surveys for environnemental and engineering studie.* Practical guide to 2D and 3D surveys.

Mari J.-L., Arens G., Chapellier D., Gaudiani P, 1998. *Géophysique de gisement et de génie civil,* Publications de l'Institut français du pétrole, Paris, Éditions Technip.

Magnin O. Bertrand Y., 2005. Guide sismique réfraction, Cahier de l'AGAP n°2, Ed. du Laboratoire central des Ponts et Chausées

McNeill J.-D., Labson V., 1991. Geophysical mapping using VLF radio fields. *In* M. N. Nabighian (ed.) Electromagnetic Methods in Applied Geophysics, Soc. Explor. Geophys., vol. 2, part B, 521-640.

McNeill J.-D., 1980a. *Electrical conductivity of soil and rocks*, Technical Notes TN-5, Geonics Limited.

McNeill J.-D., 1980b. *Electrical terrain conductivity measurement at low induction numbers,* Technical Notes TN-6, Geonics Limited.

Mériaux P., Hollier-Larousse A., Derobert X., 2003. *Reconnaissance géophysique des digues de l'Agly après la crue de novembre 1999 – Contribution à l'élaboration d'une méthode de diagnostic.* Geofcan.

Mériaux P., Royet P. et Folton C., rééd. 2004. *Surveillance, entretien et diagnostic des digues de protection contre les inondations.* Guide pratique à l'usage des gestionnaires, Paris, Cemagref Éditions.

Militzer M., Rösler R., Lösch W., 1979. Theoretical and experimental investigations for cavity research with geoelectrical resistivity methods. *Geophysical Prospecting,* vol. 27, 640-652.

Miller R.D., Pullan S.E., Waldner J.-S., Haeni F.P., 1986. Field comparison of seismic sources. *Geophysics,* vol. 51, 2067-2092.

Parasnis D.S., 1986. *Principles of applied geophysics.* Chapman and Hall, 4ᵉ éd.

Palacky G.J., 1991. Resistivity characteristics of geological targets. *In* M. N. Nabighian (ed.), *Electromagnetic Methods in Apllied Geophysics,* Soc. Explor. Geophys., vol. 1, 53-129.

Philipponat G., Zerhouni M., 1993. Interprétation de l'essai au phicomètre. *Revue française de géotechnique,* n° 65, 3-28.

Spies B.R. and Frischknecht F.C., 1991. Electromagnetic sounding, Electromagnetic methods in applied geophysics. *In* M. N. Nabighian (ed.), *Electromagnetic Methods in Applied Geophysics,* Soc. Explor. Geophys., vol. 2, part A, 285-386.

Tabbagh A., Benderitter Y., Andrieux P., Decriaud J.-P., Guerin R., 1991. VLF resisitivity maping and verticalization of the electric field, *Geophysical Prospecting,* vol. 39, 1083-1097.

Roy A., Apparao A., 1971. Depth of investigation in direct current methods, *Geophysics,* 36 (5), 943-959.

Ursat P., 1995. *Le Perméafor, un appareil bien adapté à l'analyse des fuites dans les digues en terre,* 11ᵉ Conférence européenne de méca-sols et travaux de fondation, 28 mai – 1ᵉʳ juin 1995, Copenhague.

Ursat P., 1992. Le Perméafor. Appareillage de diagraphie de perméabilité, *Bulletin de liaison des Ponts et chaussées,* vol. 178, Réf. 3641, 19-26.

West G.-F. et Macnae J.-C., 1991. Physics of the electromagnetic induction exploration method, Electromagnetic methods in applied geophysics. *In* M. N. Nabighian (ed.), *Electromagnetic Methods in Applied Geophysics,* Soc. Explor. Geophys., vol. 2, part A, 5-45.

Bibliography

PHOTO CREDITS

Introduction, p. 9
Localised zone of overtopping, repaired, on the land side of the dike on the right bank of the Agly river at Saint-Laurent-de-la-Salanque, following damage caused by flooding on 12/13/11/1999, P. Mériaux.

Chapter 2, p. 13
Dike on the Petit-Rhone, showing seepage during the flood of 8 January 1994, P. Mériaux (8/01/1994).

Chapter 3, p. 25
Typical profile of a dike in the middle reaches of the Loire, P. Mériaux.

Chapter 4, p. 39
Exploration using the Slingram method (EM 31 device), EDG.

Chapter 5, p. 91
Sounding with a light-duty dynamic penetrometer, P. Mériaux.

Chapter 6, p. 103
Geophysical exploration using RMT, P. Mériaux.

Chapter 7, p. 111
Panorama of the breach at St-Laurent-de-la-Salanque, Éric Josse (DDE 66).

Conclusion, p. 113
Downstream draining shoulder constructed after the flood of January 8 1994 on the dike along the Petit Rhone at Arles; service road built at the top of the weighted slope – the dike crest remains unsuitable for vehicles, P. Mériaux.

List
OF FIGURES

Imprimé pour vous par Books on Demand (Allemagne)

www.ingramcontent.com/pod-product-compliance
Lightning Source LLC
LaVergne TN
LVHW010819200726
843507LV00003B/644